An Introduction to Design of Piers and Wharves

J. Paul Guyer, P.E., R.A.

Editor

The Clubhouse Press
El Macero, California

CHAPTERS

(Figures, tables and formulas in this publication may at times be a little difficult to read, but they are the best available. **DO NOT PURCHASE THIS PUBLICATION IF THIS LIMITATION IS UNACCEPTABLE TO YOU.**)

CHAPTER 1
PLANNING

1. INTRODUCTION

1.1 SCOPE. This publication contains descriptions and design criteria for pier and wharf construction, including subsidiary, contiguous, and auxiliary structures. Loading details, regulations, furnishings, appurtenances, and other information are discussed when applicable. This publication provides minimum facility planning and design criteria for efficient homeporting facilities of vessels. Existing ports, facilities, and berths may not meet all criteria and may therefore, perform less efficiently, but do not necessarily require upgrade. This publication focuses on the entire homeport operation.

1.2 GENERAL FUNCTION. An important consideration that often comes up has to do with differentiation between homeports and ports of call. Basically, a homeport for a specific ship has been identified as such by the Owner. The homeport is where the ship is assigned and offers all requisite services required by the ship to include the full compliment of hotel services. In contrast, a port of call would be any port where a ship stops along the way other than its homeport, or a stop at a fueling pier, a supply pier, or a repair pier. The only real requirements for a port of call would be that it has sufficient dredge depth and that it provides secure mooring. Ship does not go cold iron in port of call and uses its organic systems. However, local determinations and justifications can warrant adding specific features at ports of call. Many of the new classes of ships have concepts of operations and special mission requirements that have resulted in making accommodations at ports in forward operating areas that ordinarily would not be required, i.e. hotel services. These are handled on a case-by-case basis and driven by operational requirements. Generally, piers and wharves provide:

- Berths with sufficient dredge depths for vessels.
- Secure mooring for vessels berths.

- Transfer points for cargo and/or passengers between water carriers and land transport.
- Facilities for fitting-out, refit or repair; and specialized functions.

1.3 FUNCTIONAL CATEGORIES. Piers and wharves are grouped into four (4) primary types as follows:

1.3.1 TYPE I – FUELING, AMMUNITION, AND SUPPLY.

1.3.1.1 FUELING. These are dedicated piers and wharves equipped with facilities for off-loading fuel from ship to storage and for fueling ships from storage.

1.3.1.2 AMMUNITION. These are dedicated military piers and wharves used for discharging ammunition for storage and for loading ammunition on outgoing ships.

1.3.1.3 SUPPLY. Supply piers and wharves are used primarily for the transfer of cargo between ships and shore facilities. Provide standard gage railroad tracks when supplies will be brought in by rail.

1.3.2 TYPE II – GENERAL PURPOSE.

1.3.2.1 BERTHING. These are general-purpose piers and wharves used primarily for mooring ships. Furthermore, berthing facilities may be active, as when ships are berthed for relatively short times and are ready to put to sea on short notice, and inactive, as when they are berthed for long periods in a reserve status. Activities that typically take place on berthing piers and wharves are personnel transfer, maintenance, crew training, cargo transfer, light repair work, and waste handling. Under some circumstances, fueling and weapons system testing may also be carried out in these facilities.

1.3.3 TYPE III – REPAIR. An important consideration for repair piers is the need to provide heavy weather mooring capability. This includes properly sized and spaced storm bollards and a compliment of heavy weather mooring lines. This consideration is predicated upon the fact that ships under repair may not be able to get underway during a heavy weather event.

1.3.3.1 REPAIR. Repair piers and wharves are constructed and equipped to permit overhaul of ships and portions of a hull above the waterline. These structures are generally equipped with portal cranes or designed to accommodate heavy mobile cranes.

1.3.3.2 FITTING-OUT OR REFIT. Piers and wharves for fitting-out are very similar to those used for repair purposes, providing approximately the same facilities. However, fitting-out military piers and wharves will have, in addition to light and heavy portal cranes, large fixed tower cranes for handling guns, turrets, engines, and heavy armor.

1.3.3.3 FLOATING DRYDOCKS. Piers and wharves for floating drydocks are constructed and equipped to permit overhaul of ships above and below the waterline. Some floating drydocks have portal cranes on tracks on the wingwalls and some floating drydocks use cranes from the pier side. The dredge depth at these facilities must accommodate the floating drydock when submerged. Floating Drydocks are normally moored using two or more vertical spuds that maintain the horizontal position of the dock throughout its full range of vertical movement from fully submerged to fully dewatered. The pier/wharf structure must be designed to accommodate mooring spud placement and loading. The pier or wharf layout should also consider personnel, material, and vehicle access to the drydock pontoon deck when the drydock is in the raised (dewatered) position.

1.3.4 TYPE IV – SPECIALIZED.

1.3.4.1 MAGNETIC TREATMENT AND ELECTROMAGNETIC ROLL PIERS. These are piers that moor ships over an array of underwater instruments and large-area cable solenoids used specifically for removing and/or modifying the magnetic signature characteristics of military surface vessels and submarines, as well as calibrating the on-board degaussing systems of mine countermeasure vessels. Magnetic Treatment facility designs vary using slips, T-piers, or single piers depending upon location and requirement. Magnetic Treatment piers designed to accommodate surface vessels are typically configured as T-shapes; whereas, submarines and mine countermeasure vessels are typically treated at drive-in piers built in parallel configurations.

1.3.4.2 TRAINING, SMALL CRAFT, AND SPECIALIZED VESSELS. These piers and wharves are typically light structures designed for specific but limited functions. Specific requirements are usually provided by the Owner.

1.4 FLEXIBILITY OF BERTHS. Typically, piers and wharves are designed to provide space, utility service, and other supporting facilities for specific incoming or homeported ships. However, berthing plans and classes of ships berthed change with time. While it is not economically feasible to develop a single facility to accommodate and service all classes of ships presently known, design the facility with a certain amount of flexibility built in for anticipated future changes in the functional requirements. This is especially true for berthing piers and wharves that will be used to accommodate different classes of ships as well as support a variety of new operations.

1.5 APPURTENANCES AND FEATURES. The following is a range of appurtenances and features that may be required for piers and wharves.

- Mooring devices to safely secure the ship.
- Fender systems, camels and separators.
- Hotel and ship service utilities.
- Communications.

- Cranes and crane trackage.
- Access facilities for railroad cars and trucks.
- Waste handling facilities.
- Cargo handling equipment.
- Covered and open storage spaces for cargo.
- Support building, tool shed, office space, and control rooms.
- Lighting poles and equipment.
- Security systems.
- Firefighting equipment.
- Emergency medical facilities.
- Access structures and facilities.
- Fueling facilities.
- Safety Ladders.
- Life-Safety Rings.

2. FACILITY PLANNING

2.1 LOCATION AND ORIENTATION.

2.1.1 GENERAL The location and alignment of piers and wharves in a harbor should consider factors such as:

- maneuverability,
- required quayage,
- harbor line restrictions,
- geotechnical conditions,
- isolation requirements,
- prevailing wind and current directions,
- clearance between moored or passing vessels,
- project depth,
- shoaling patterns,
- environmental permit restrictions,
- port regulations and
- landside access/proximity.

2.1.2 ORIENTATION FOR ENVIRONMENTAL CONDITIONS. To the extent practicable, orient piers and wharves so that a moored ship is headed into the direction of the prevailing winds and currents. Thus, the forces induced on mooring lines by these conditions would be kept to a minimum. If such an arrangement is not feasible, consider an orientation in which the wind or current holds the ship off the facility, although do not overlook the difficulty in mooring a ship under such conditions. In locations where criteria for both wind and current cannot be met, orient the berth parallel to the direction of the more severe condition. At locations exposed to waves and swell, locate the facility so that a moored ship is headed into the wave or swell front. If planning criteria dictate that a pier or wharf be oriented so that a moored ship is positioned broadside to the prevailing winds, currents, or waves, consider breast-off buoys to keep the ship off the facility and diminish the possibility of damage to the structure and ship. At oil storage terminals located in areas where meteorological and hydrological conditions are severe, consider using a single point mooring which allows a moored tanker to swing freely when acted upon by winds, waves, and currents from varying directions.

2.1.2.1 PIER ORIENTATION. A pier is oriented either perpendicular to or at an angle with the shore. There are generally slips on both sides, although there are instances where only one side has a slip because of site conditions or because there is no need for additional berthing. Piers may be more desirable than wharves when there is limited space available because both sides of a pier may be used for mooring ships. When both sides of a moored ship need to be accessed, two parallel piers with a slip in between may be preferred. Magnetic Treatment and Electromagnetic Roll piers usually require a magnetic north/south orientation, irrespective of other considerations

2.1.2.2 WHARF ORIENTATION. A wharf is a structure oriented approximately parallel to the shore. Ships can only be moored at the offshore face of a wharf. When water depths close to shore are not adequate to accommodate deep draft ships, the wharf, consisting of a platform on piles, is located offshore in deep water and is connected to shore at one or more points by pile-supported trestles, usually at right angles to the wharf. If the trestle is located at the center of the wharf, the structure is referred to as a T-type wharf; if the trestle is located at an end, the facility is known as an L-type wharf; if trestles are located at both ends, the wharf is called a U-type wharf. Ships may be berthed on both sides of a T- or L-wharf. When the offshore wharf is used for transfer of bulk liquid cargo from the unloading platform to shore via submarine pipelines, the structure is referred to as an island wharf. A trestle from the offshore wharf to shore is not provided and both sides of the island wharf may be used for mooring ships.

Launches are used for wharf access. Where a U-shaped berth is formed by a cut into land by two approximately parallel wharves, this may be referred to as a slip. For examples of pier and wharf types, see Figures 2-1 and 2-2. For general cargo, supply, and container terminals, a wharf structure, connected to upland shore area for its full length, is preferred because such an arrangement is more adaptable to loop rail and highway connections and the distance from wharf apron to transit sheds and open storage areas is shorter.

2.1.3 VESSEL INGRESS AND EGRESS. On occasion, a moored vessel is required to make a hasty departure from its berth and head out to sea. Accordingly, when planning a pier or wharf, consider providing adequate turning area so that a ship can be turned before it is docked, and moored with a heading that will permit a convenient and rapid departure.

2.1.3.1 DISASTER CONTROL AND EMERGENCY PLANS. In an emergency, tugs may not be available. Therefore, the slip, berth, basin, and channels should allow the ship to get underway without assistance. There will be cases where a ship will not be able to leave prior to heavy weather. General-purpose berths are normally designed for 64-knot (75-mph (32.9 m/s)) winds. Winds in excess of 64 knots (75 mph (32.9 m/s)), such as may be seen at repair berths, require special considerations. Design facility systems for continuous operation in the event of a power outage, in other words, pier/wharf remains operational with exception of having shore power and lighting.

2.1.3.2 HARBOR ELEMENTS.

2.1.4 WATER DEPTH. At locations where the required depth of water is available close to shore and the harbor bottom slopes steeply out to deeper water, it may not be economical to build deep-water foundations for a pier; consider a wharf structure. At locations where water depths are shallow and extensive dredging would be required to provide the required depth of water close to shore, consider locating the facility offshore, in deeper water, by utilizing a T-, L-, or U-type wharf.

2.1.5 DOLPHINS. These are small independent platforms or groups of piles used by themselves or in conjunction with a pier or wharf for specialized purposes. A mooring dolphin is sometimes used at the offshore end of a pier or both ends of a wharf to tie up the bow or stern line of a ship at a more favorable angle. Mooring dolphins are usually accessed by a catwalk, as illustrated in Figure 2-1, and are provided with a bollard or capstan. Breasting dolphins are sometimes used for roll-on/roll-off facilities and at fueling terminals where a full-length pier or wharf is not required. They may also be used as part of the fender system. A turning dolphin is an isolated structure used solely for guiding the ships into a berth or away from known obstructions. Occasionally, a mooring dolphin may also be designed to function as a turning dolphin. Approach dolphins are used where the end of a pier or ends of a slip require protection from incoming ships.

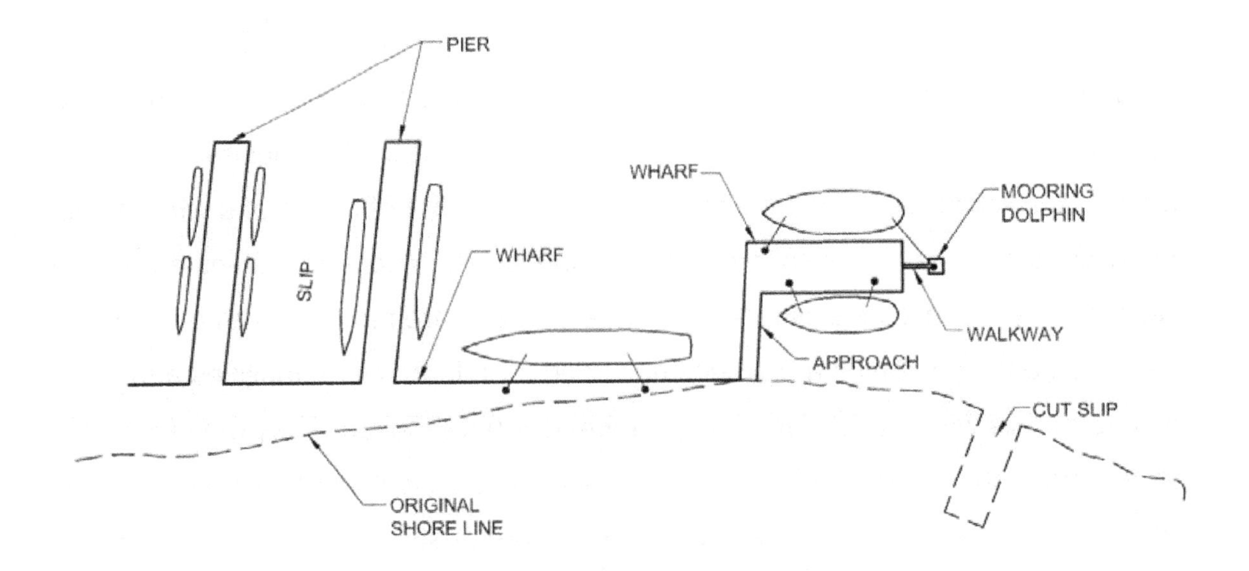

(A) PIER AND WHARF

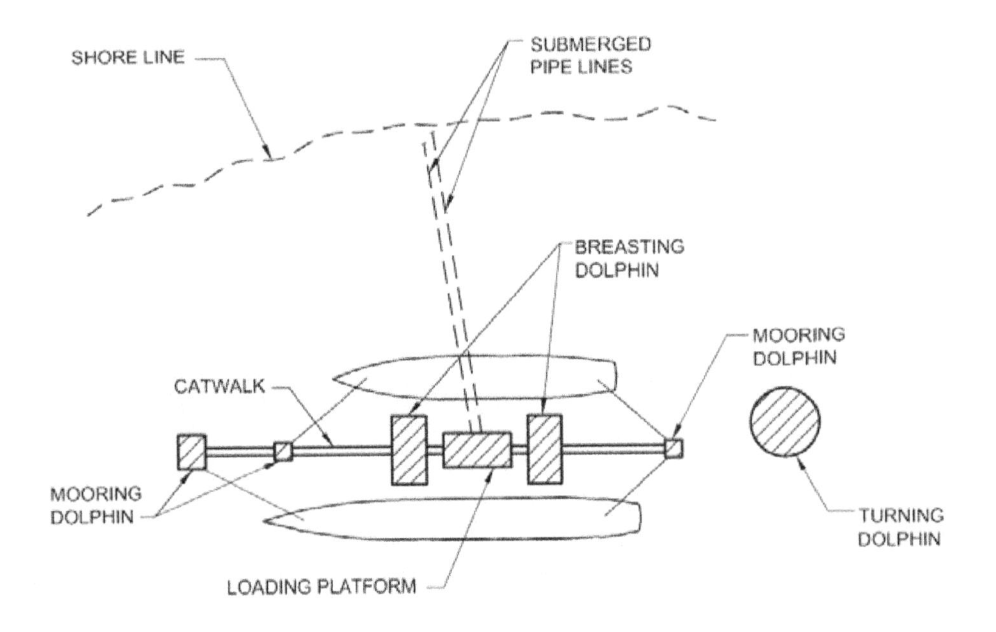

(B) ISLAND WHARF

Figure 2-1

Pier and wharf types

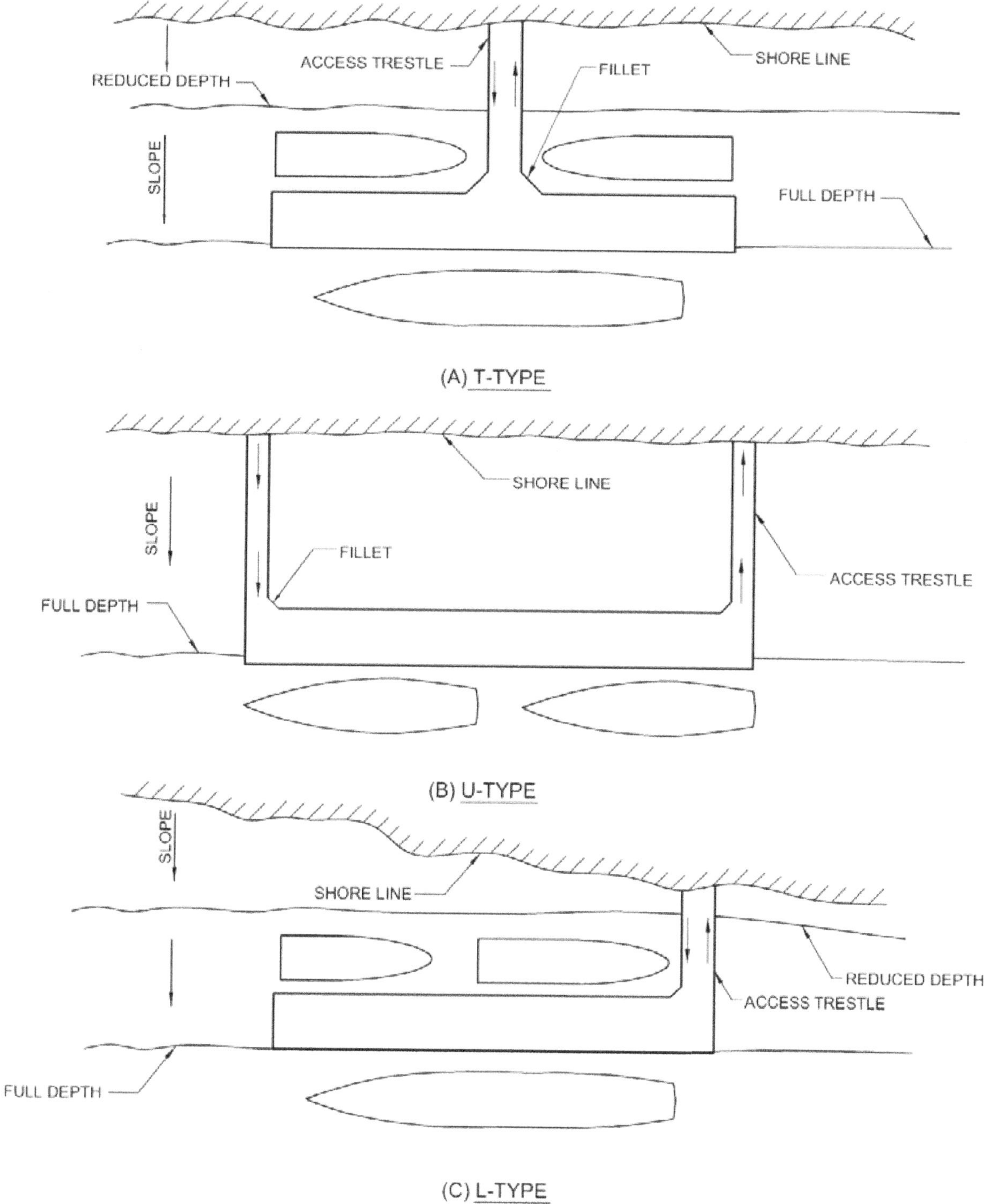

Figure 2-2

Wharf types

2.2 COORDINATION OF REPAIR.

2.2.1 GENERAL. Maintenance activities will generally be conducted at a repair berth. Coordinate capability of local ship repair facilities and salvage operations with the Owner. The following facilities should be available within a reasonable distance from the support facility homeport.

- A Ship Maintenance Facility (SMF) housing the machine tools, industrial processes and work functions necessary to perform non-radiological depot level maintenance on ship propulsion plants.
- A Maintenance Support Facility (MSF) housing both administrative and technical staff offices supporting ship propulsion plant maintenance, as well as central area for receiving, inspecting, shipping, and storing hazardous/mixed waste materials and maintenance materials, and controlled radiological tank storage.
- A Controlled Industrial Facility (CIF) or Radiological Work Facility used for the inspection, modification, and repair of radiological controlled equipment and components associated with nuclear propulsion plants. It also provides facilities and equipment for the treatment, reclamation, and packaging for disposal of radiologically controlled liquids and solids. It includes non-radiologically controlled spaces for administration and other support functions.
- Drydock facilities for various classes of ships.

2.3. OVERALL DIMENSIONS AND CLEARANCES

2.3.1 GENERAL. The overall dimensions and clearances required for piers and wharves are dependent on characteristics of the ships to be berthed and the support services provided. Supporting items may include:

- Provide fender / camel systems as required by Owner.
- A minimum of 5 acres of laydown area in addition to pier/wharf space is desirable. The laydown area should be within ½ mile of the pier or wharf.

- Brows and Platforms are usually placed at ship's designated entry/egress points to the main deck. Brow design length will be based on camel design and resultant standoff distance.
- Design facilities (ramps, landings, railings) in accordance with American with Disabilities Act (ADA) of 1990.
- Design general warehouse space to be accessible to large trucks and handling equipment.
- Do not permit pier interferences such as utilities and deck appurtenances.
- Provide parking for ship's vehicles in proximity to entrance to the pier or wharf.
- Provide sufficient security for ships on the pier.
- Provide safety equipment.

2.3.2 PIER AND WHARF LENGTH.

2.3.2.1 SINGLE BERTH. The length of pier or wharf should equal the overall length of the largest ship to be accommodated, plus an allowance of 50 ft at each end of the ship.

2.3.2.2 MULTIPLE BERTHS. The length of a pier or wharf should equal the total overall length of the largest ships simultaneously accommodated, plus clear distance allowances of 100 ft between ships and 50 ft beyond outermost moored ships.

2.3.2.3 CONTAINER AND RO/RO BERTHS. The length of berths used for container or RO/RO berths should account for the requirements of the container cranes or special ramps. Where shipboard ramps are used, provide adequate berth length to allow for efficient vehicle maneuvering.

2.3.3 PIER AND WHARF WIDTH. Pier width refers to the net operating width of the structure, exclusive of fender systems, curbs, and dedicated utility corridors. This definition also holds for U-, L-, and T-type wharves. However, with reference to wharves, the width should be the dimension to a building, roadway, or other identifiable obstruction. Review with specific functional requirements of the individual installation in mind before a final selection is made. Functional requirements include space for: cargo loading operations, line handling, ship maintenance, maintenance of utilities and layout of cables and hoses, solid waste collection, brows and platforms, crane operation, and other operations. For crane operation, consider crane outriggers, tail swing of crane counterweights, and overhang of vessels. Also, these dimensions should not be less than the widths determined by geotechnical and structural considerations. Factors to be considered in the determination of pier and wharf width are discussed below.

2.3.3.1 UTILITIES. One of the primary functions of a pier or wharf is to provide connections for utilities from ship to shore. Fixed utility terminals are usually provided close to the edge of the pier or wharf along the bullrail. Flexible hoses and cables are then connected to these terminals and to the ship. Depending upon the type of utility hoods, the terminals, hoses, and cables may require 10 to 15 ft of space along the edge that cannot be utilized for any other purpose. Consider the types of utility hoods that require additional edge space for cable and hose laydown. On single deck piers, substations are typically mounted on the deck, which require an additional 25 ft of pier width. Double deck piers are used where the width of the berth area is constrained by adjacent facilities or other limitations. This configuration allows the utility enclosures and the associated hoses, cables and maintenance activities to be segregated from the operational areas and allow crane operations closer to the edge of the pier or wharf.

2.3.3.2. BERTHS ON ONE SIDE. For wharves or piers with berths on one side, the minimum width is 66 ft comprised of: 16 ft bollards and utilities, 35 ft mobile crane ops, and 15 ft fire lane.

2.3.3.3 BERTHS ON BOTH SIDES. For Single deck pier with berths on both sides, the minimum width is 117 ft. comprised of: 32 ft bollards and utilities, 70 ft mobile crane ops/loading area, and 15 ft fire lane.

2.3.3.4 DOUBLE DECK PIER WITH BERTHS ON BOTH SIDES. Minimum width is 93-ft comprised of: 8 ft bollards, 70 ft mobile crane ops/loading area, and 15 ft fire lane. A double deck pier provides: clear unobstructed pier to ship interface; isolation of operations deck services from lower deck utilities services (i.e. substation located on lower deck); reduced offset requirements for mobile crane operation (thereby reducing the requirement for floating cranes); higher main deck, improving mooring line angles and lessening need for brow platforms.

2.3.4 MOBILE CRANE OPERATION. With the exception of magnetic treatment/ electromagnetic roll and fueling piers where a lighter duty mobile crane and/or forklift truck is sufficient, piers and wharves are subject to frequent usage by mobile cranes, forklifts, and straddle carriers. Typically, the cranes will be used to lift light loads (5 to 10 tons) but at a longer reach. This requires a high-capacity crane. If the crane operations are not allowable because of utility trenches and trenches with light-duty covers, such areas should be clearly marked and separated by a raised curb to prevent accidental usage. Typically, mobile crane operators want to get as close as possible to the edge of the pier or wharf to reduce the reach. However, the edges of piers and wharves are also the best places for locating utility trenches and utility trenches. This conflict can be resolved by either designing all utility covers to the high concentrated load from the mobile crane or by allowing crane operations in discrete and dedicated spaces along the edges. Weight-handling equipment requires maneuvering and turnaround space on the deck for effective operation. If possible the deck space should be planned to allow mobile cranes to be backed up perpendicular to the bullrail. This permits the maximum load/reach combination. Make allowance for tail-swing of crane counter-weight.

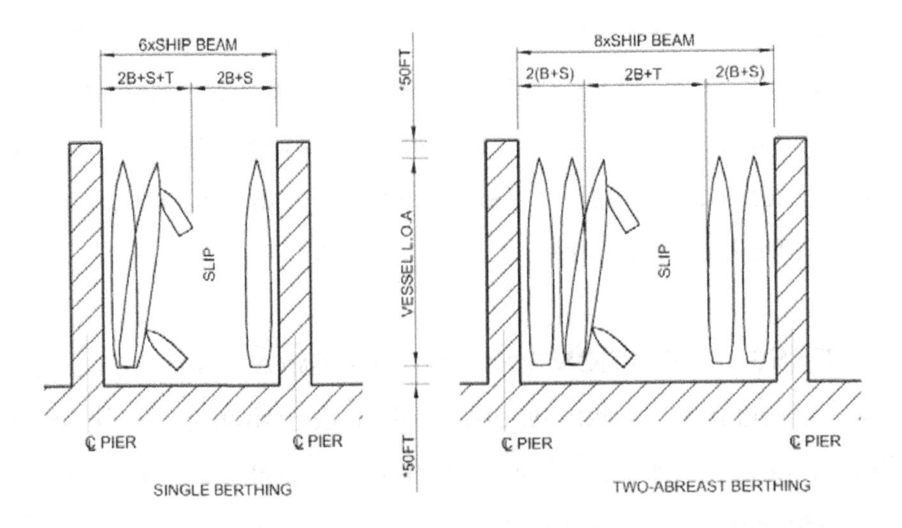

SINGLE BERTH PIERS

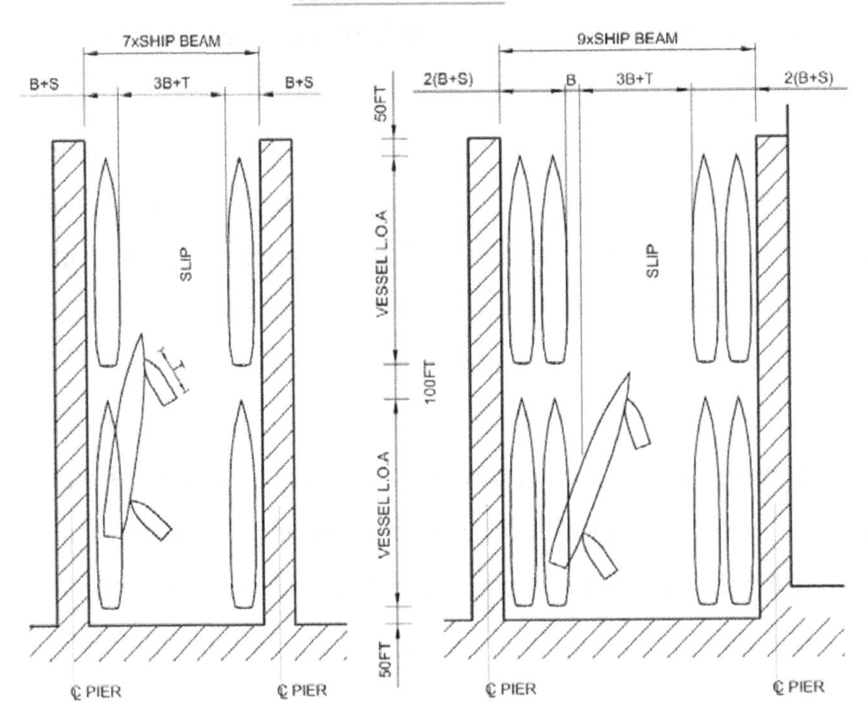

MULTIPLE BERTH PIERS

Figure 2-3

Length and width of slip

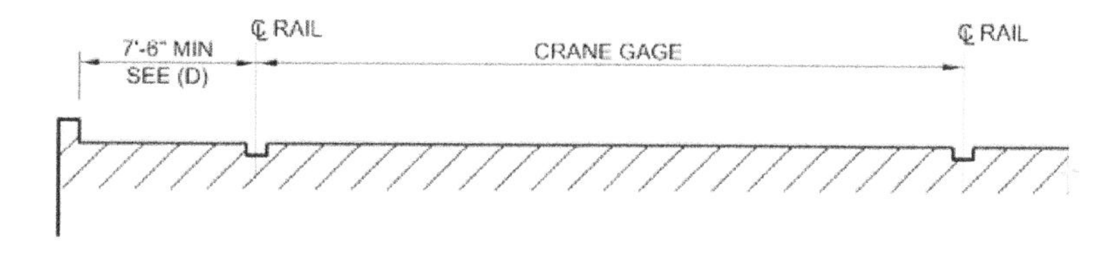

(A) CRANE RAIL ONLY

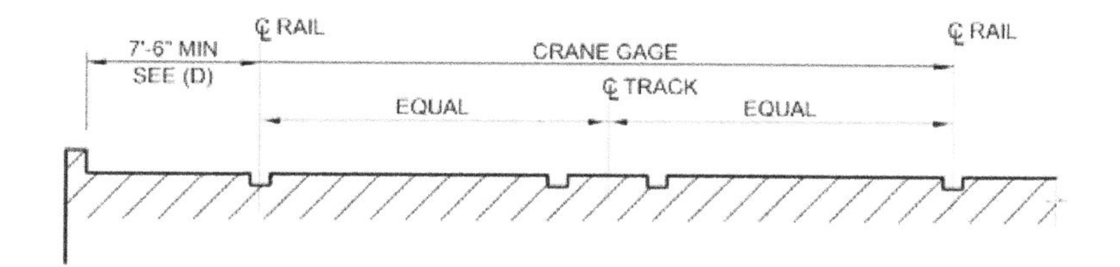

(B) CRANE RAIL AND SINGLE TRACK RAILROAD

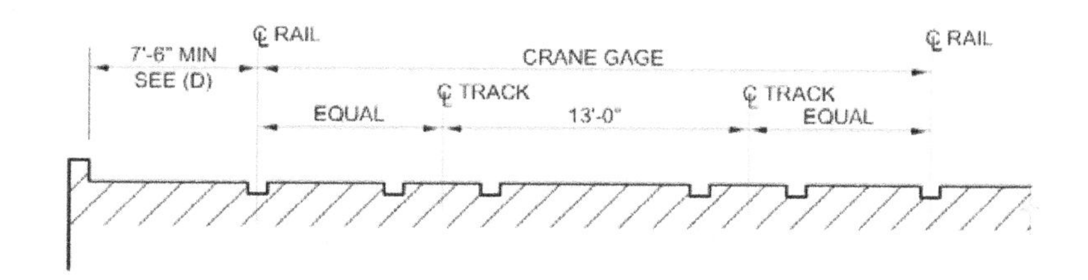

(C) CRANE RAIL AND DOUBLE TRACK RAILROAD

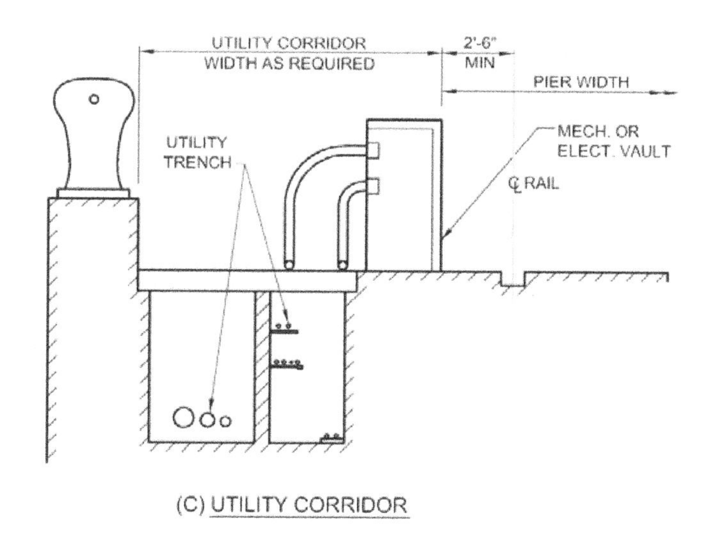

(C) UTILITY CORRIDOR

Figure 2-4

Location of crane rails, railroad tracks and utilities

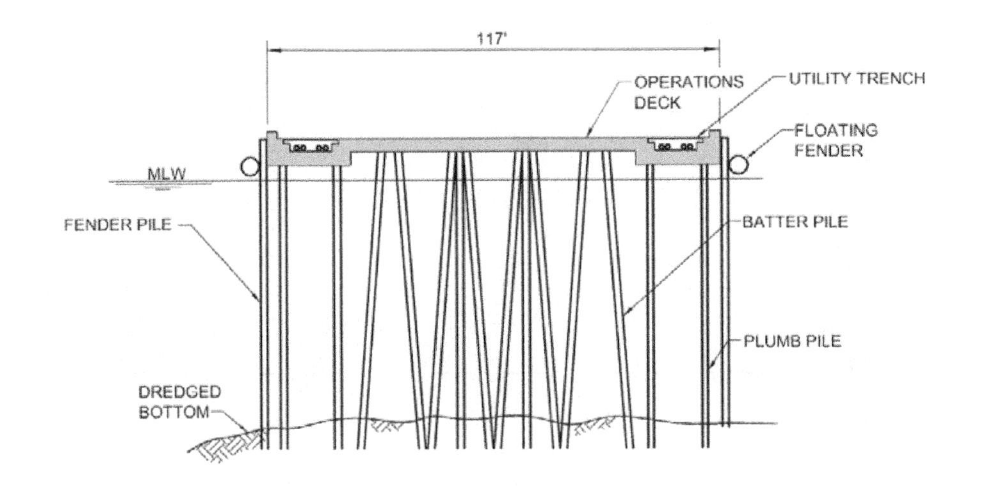

(A) SECTION VIEW

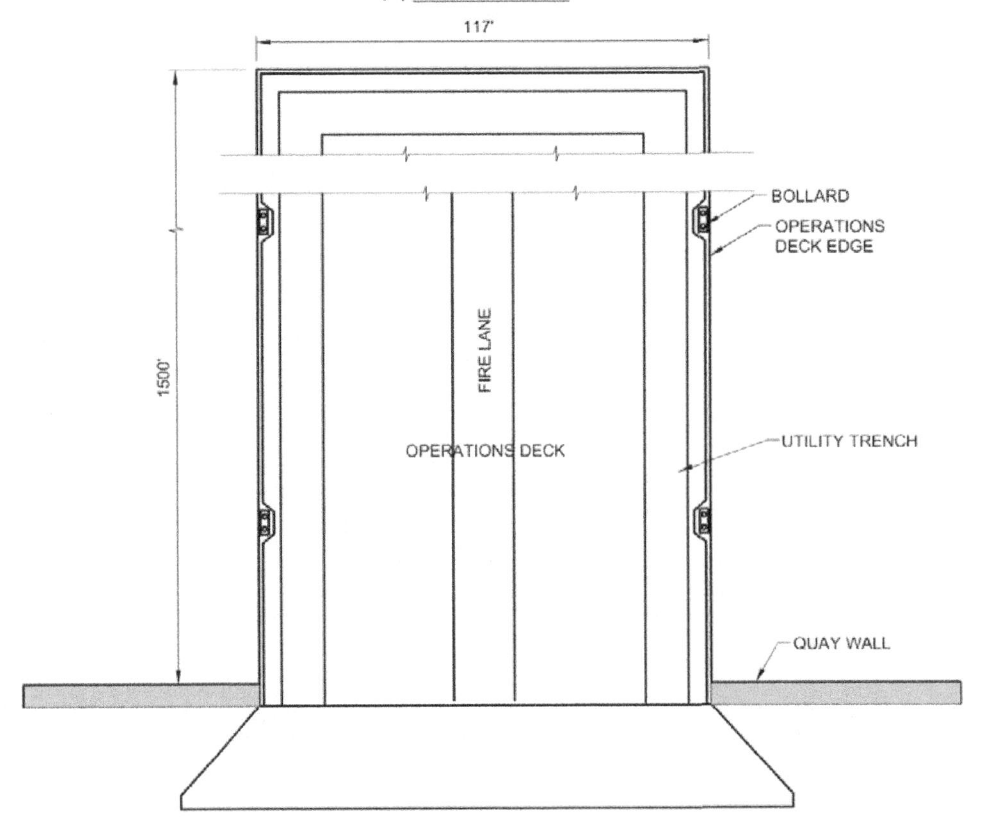

(B) PLAN VIEW

Figure 2-5

Single deck pier

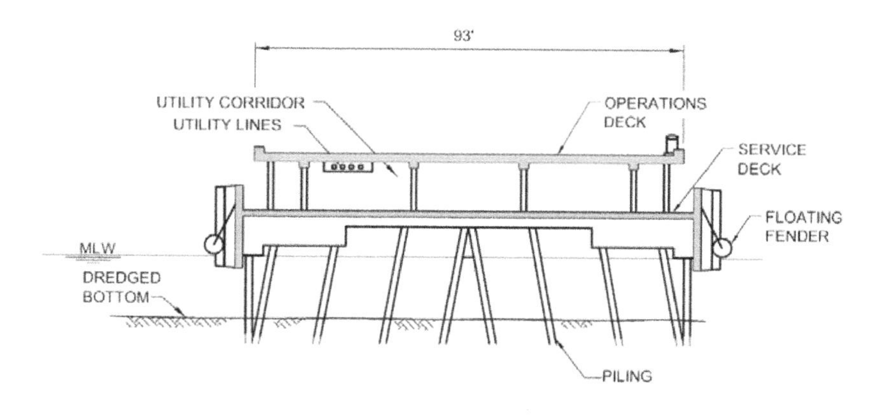

(A) SECTION VIEW

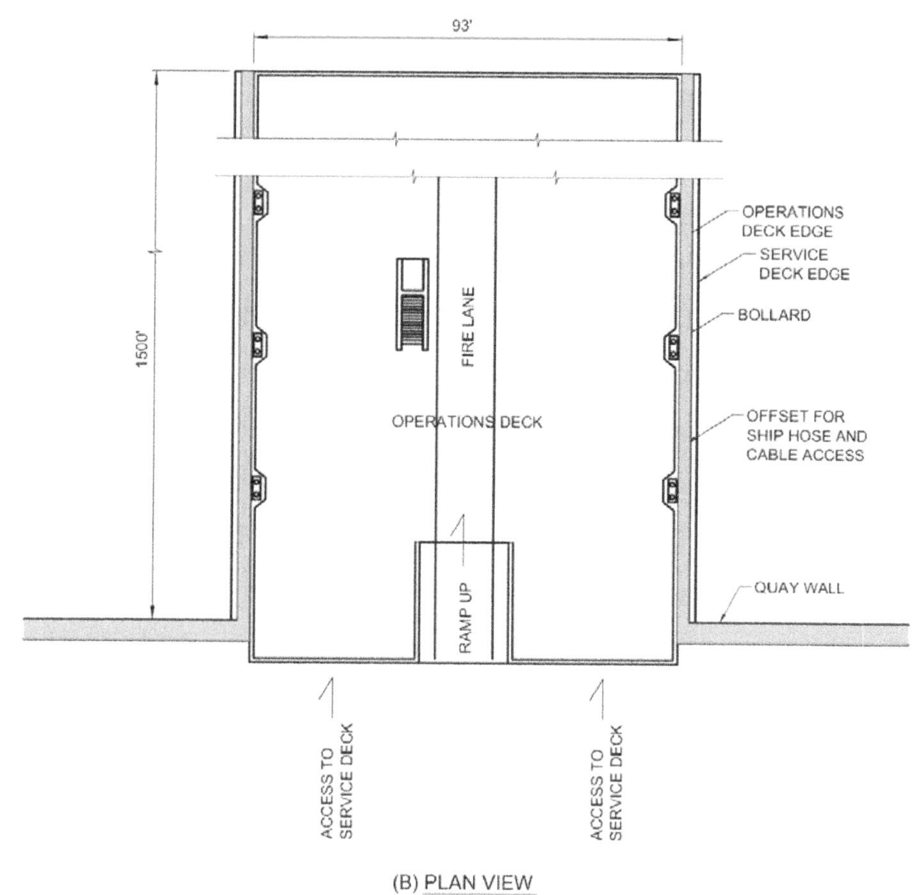

(B) PLAN VIEW

Figure 2-6

Double deck pier

2.3.5 RAILROAD AND CRANE TRACKAGE. The number of railroad and crane tracks required and type of weight-handling equipment furnished on piers and wharves are dependent on the type of function, ships to be accommodated, amount of cargo to be handled, and rate of cargo transfer. Specific service requirements of the individual installation should be evaluated in conjunction with the following considerations: Rail mounted cranes are often needed for loadout in repair, fitting out, and refit facilities. Width requirements depend on equipment selected. The use of wide gage crane service at repair, fitting out, ammunition and supply piers, and wharves should be considered. A rail gage of 40 ft for new cranes, except at container terminals or where it is necessary to conform to gages of existing tracks. When cranes are furnished, the distance from the waterside crane rail to the edge of the pier or wharf should be adequate to provide clearance for bollards, cleats, capstans, pits housing outlets for ship services, crane power conductors, and other equipment. Some electric powered gantry cranes may require either open or covered (panzer-belts) cable trenches in the pier or wharf deck for the power conductors. Where locomotive cranes are used on piers and wharves, the distances between tracks and curbs should be increased to accommodate the tail swings of the crane. Do not consider railroad trackage for use on berthing piers and wharves (both active or inactive), except at stations where most cargo is received by rail, one or more tracks may be considered for use on active berthing piers. When there are existing railroad networks at the station, tracks should be considered for installation on repair, fitting out, ammunition and supply piers and wharves. When trackage is required along aprons of piers and wharves, at least two tracks should be provided so that one track may be used as a running track when the other track is occupied. Track gage should conform to gage of existing trackage on adjacent piers and wharves to avoid creation of "captive" cranes. Except where local conditions require otherwise, standard gage should be used for trackage. Width of piers and wharves should be adequate to allow passing of trains and forklift trucks (or other material-handling equipment). Make allowances for stored cargo and other obstructions. Where sponsons or flight decks or other ship types with large deck overhangs are anticipated to be berthed, locate the crane rail so that all parts of the crane will clear the deck overhang. Railroad and crane trackage should not be considered for use on piers and wharves used primarily for fueling (petroleum) operations. When railroad and crane trackage is required on piers and wharves, the spacings shown on Figure 2-4 may be used as a guide.

2.3.6 TRUCKS AND OTHER VEHICLES. A variety of service trucks and vehicles can be expected to use piers and wharves for moving personnel, cargo, containers, and supplies to and from the ships. The width provided must take into account operation and maneuvering of such vehicles. Turnaround areas should be provided.

2.3.7 SHEDS AND BUILDINGS. Pier and wharf deck is usually too expensive an area for storage sheds, which should therefore be located on land to be cost-effective. Storage sheds and buildings of any kind should be kept off piers and wharves unless their location can be justified by security considerations. Transit sheds may be considered on piers and wharves where a suitable upland area is not available. When used on a pier, the transit shed should be located along the centerline with clear aprons on both sides consistent with the requirements set forth herein but not less than 20 ft (6.1m) or more wide. On wharves, transit sheds and support buildings should be located on the land side edge with a clear apron toward the waterside. In general, support buildings on piers and wharves should be kept as small as feasible and located away from high-activity areas for least interference.

2.3.8 MOVABLE CONTAINERS AND TRAILERS. During active berthing of ships, various containers of different sizes are temporarily or permanently located on pier deck to support the operations. These include shipyard toolboxes, garbage dumpsters, training trailers, and supply trucks. Adequate deck space should be available for locating and accessing these containers and trailers.

2.3.9 FIRE LANE. Retain and mark, with a painted yellow line, a 15-foot-wide unobstructed fire lane. Provide a marked (with dashed, painted yellow lines) 2.5-foot wide "dual use" buffer on both sides of the 15-foot-wide unobstructed fire lane. Local enforcement would keep this area clear enabling a total available fire lane width of 20 feet. However, encroachment into the "dual use" buffer by cranes is acceptable. Any other encroachment into the "dual use" buffer will be handled on a case-by-case basis. For wharves, provide a 20-foot-wide unobstructed fire lane immediately adjacent to the operating area. These requirements need not be applied to small craft or yard craft piers.

2.3.10 FUEL-HANDLING EQUIPMENT. At specified berths, stationary fuel-handling equipment consisting of self-adjusting loading arms is often furnished to offload fuel products from tankers to onshore storage facilities. Pier or wharf width requirements depend on equipment selected and facilities furnished.

4. SLIP.

04.1 GENERAL CONSIDERATIONS. The clear distance between piers, or slip width, should be adequate to permit the safe docking and undocking of the maximum size ships that are to be accommodated in the slip. The size of a slip should also permit the safe maneuvering and working of tugboats, barges, lighters, and floating cranes. At multiple berth piers, where ships are docked either one per berth, two abreast per berth, or more, sufficient clearance should be available to permit the docking and undocking of ships at the inboard berth without interfering with ships at the outboard berth. Because the size of a slip is affected by docking and undocking maneuvers, consideration should be given to the advice of local pilots who are familiar with the ships to be handled and with prevailing environmental conditions such as winds, waves, swells, and currents. Slip width is also influenced by the size and location of camels/separators used between ship and structure and between ships. The width should be reviewed with specific function al requirements of the individual installation before a final determination is made.

2.4.2 MINIMUM SLIP LENGTH - 1350-1500 ft. min. This range in slip length is based on berthing requirements.

2.4.3 MINIMUM WIDTH OF SLIP FOR ACTIVE BERTHING. Minimum width should be the greater of the two dimensions shown. Additionally, the width should not be less than 300 ft. The recommended criteria are applicable only if ships are turned outside the slip area. Refer to SCDB for the beam of typical ship types. The requirements apply where ships are berthed on both sides of a slip. Where ships are berthed on only one side of a slip, the width may be reduced. When more than two abreast berthing is employed, the width of slip should be increased by one ship beam for each additional ship added in order to maintain adequate clearances between moored ships during berthing and unberthing maneuvers. Thus, for three-abreast berthing on both sides of a slip, the slip width for single-berth piers would be equal to 10 times ship beam and the slip width for multiple-berth piers would be equal to 11 times ship beam.

2.4.4 MINIMUM WIDTH OF SLIP FOR INACTIVE BERTHING. At slips containing inactive berths where ships are stored for long periods of time on inactive status in nests of two, three, or more, clear distances between moored ships and slip width may be reduced by one or two ship beams to reflect the reduction in the frequency of berthing maneuvers and the decrease in activities of small boats and floating equipment.

2.4.5 WATER DEPTH IN SLIPS. Information on required water depth in slips and at berths shall be provided by the Owner.

2.5 PIER AND WHARF DECK ELEVATION. Set deck elevation as high as possible for surface ship berthing based on the following considerations:

2.5.1 OVERTOPPING. To avoid overflow, deck elevations should be set at a distance above mean higher high water (MHHW) level equal to two thirds of the maximum wave height, if any, plus a freeboard of at least 3 ft. Bottom elevation of deck slab should be kept at least 1 ft above extreme high water (EHW) level. Where deck elevation selected would result in pile caps or beams being submerged partially or fully, consideration should be given to protecting the reinforcing from corroding.

2.5.2 SHIP FREEBOARD. Consideration should be given to the varying conditions of ship freeboard in relation to the use of brows and the operation of loading equipment such as conveyors, cranes, loading arms, and other material handling-equipment. Fully loaded ships at mean lower low water (MLLW) level and lightly loaded ships at MHHW level should be considered for evaluating the operation of such equipment.

2.5.3 UTILITIES. Deck elevations should be set high enough above MHHW levels to allow for adequate gradients in drainage piping as well as to prevent flooding of utilities/trenches that are located below the pier deck.

2.5.4 DECK ELEVATION FOR BERTHS. Deck elevation must not conflict with elevators. Use properly designed camels and pier fendering to provide sufficient standoff to prevent interference.

2.5.5 ADJACENT LAND. If possible, set deck elevation as close as possible to grade of the adjacent land for smooth access of mobile cranes, service vehicles, personnel vehicles, and railroad. Ramps may be used to access the deck set higher or lower than adjacent land. A maximum gradient of 15 percent may be used for such ramps when railroad access is not provided. Consideration should be made to reduce grade to 6 to 8 percent, comparable to state DOT requirements, in cold regions where snow or ice can be a problem. Ramps for pedestrian access should have a gradient less than 12 horizontal to 1 vertical, with 5 ft minimum landings for every 30 in of rise to conform to American with Disabilities Act (ADA) of 1990. Vertical curves should be large enough so that long wheelbase or long overhang vehicles do not or drag. Where track mounted cranes are specified, all the deck areas serviced by the crane should be kept at the same elevation.

2.5.6 SPECIAL SITUATIONS. For double deck piers or wharves and in situations where a sloping deck is contemplated (for gravity flow of sewer lines), all the above considerations should be evaluated.

2.4 UTILITIES.

2.4.1 GENERAL. Usually, utility connection points (hoods, vaults, or mounds) are located and spaced along the pier or wharf edge to be as close as possible to the ships' utility terminals in the assumed berthing position. The connection points should be planned and located to accommodate reasonable future changes in berthing plan or in the type of ships served. Typical hotel services are: potable water, non-potable/saltwater, CHT, oily waste/waste oil, compressed air, steam, telecommunications, and shore power. All utility lines should be kept where they can be conveniently accessed from above deck. Typical utility configurations on piers and wharves are described below:

2.4.1.1 UTILITY TRENCHES. These are basically protected trenches running along the waterside edge of a pier or wharf accessed by removable covers from the top. In a pier, the lines can go along one edge all the way to the end and be "looped" to the other edge back to land. In a wharf, the lines can be supplied and returned through smaller lateral "trenches." Where the number and size of lines is large enough, a utility tunnel or gallery can be utilized with access from the top or side. Where a fuel line is provided, it should be kept in a separate trench for containment of leaks.

2.4.1.2 DOUBLE DECK PIER. Utility lines are isolated on the lower deck and can be easily accessed. The upper deck is thus clear of all utility lines and terminations and is free for operations.

2.4.1.3 BALLASTED DECK. This concept consists of a sloping deck filled with 1.5 ft to 3 ft of crushed rock ballast, which provides a convenient medium to bury the utility lines and crane or rail trackage. The ballast is topped with concrete or asphalt paving, which will provide a firm-working surface for operations. The paving and ballast can be removed to access the utility lines. Concrete pavers have been used successfully for paving ballasted decks and provide improved access to utilities buried in the wharf ballast. Future changes in utilities and trackage can similarly be accommodated. Also, the ballast helps to distribute concentrated load to the deck slab, thus allowing heavier crane outrigger loads.

2.5 LIGHTING. Pier lighting is needed for security, safety, and operations. Design: interior and exterior lighting and controls and as follows:

2.5.1 TOPSIDE LIGHTING. Instances of failures of high mast lighting light ring assembly failures have been reported. It is imperative that a registered structural engineer review all contract documents (plans and specifications) and construction contract submittals specifically related to high mast light ring assemblies. This is to insure that design of light ring assembly provides for safe lowering, raising, and locking/securing into position. /1/

2.5.2 LOWER DECK LIGHTING. Per Owner requirements.

2.5.3 UNDERDECK LIGHTING. Per Owner requirements.

2.6 SECURITY. Per Owner requirements.

2.7 LANDSIDE APPROACHES

2.7.1 FUNCTION. Approaches are required to provide access from shore to piers and wharves located offshore. Usually, the approach is oriented at right angles to the shoreline. Except in special situations, approaches should consist of open-type trestle structures that minimize impediments to water flow and disturbances to the characteristics and ecology of the shoreline. The number, width, and orientation of approaches should consider the volume of traffic flow, circulation of traffic, existing roads on shore side, fire lane requirements, and interruption of service due to accidental collision damage to the approach. As approaches are also used to route utilities to the pier or wharf, the width of approaches will be further influenced by the space requirements of the utility lines being carried. Vehicle and pedestrian approach can usually be combined on the same structure. However, where a large number of personnel are anticipated to access the facility, a separate pedestrian approach should be considered.

2.7.2 ROADWAY WIDTH. Criteria is provided below:

- For infrequently accessed facilities (such as magnetic treatment/electromagnetic roll piers), the approach roadway should have a minimum width of 10 ft curb to curb for one-lane vehicular traffic.
- For fueling piers and wharves, the approach should have a minimum width of 15 ft curb to curb for clear access of emergency vehicles.
- For all other functional types, a two-way 24 ft wide curb-to-curb roadway should be provided. If two separate one-way approaches are provided for a pier or wharf for incoming and outgoing traffic, each of them may be 12 ft wide curb to curb. In any case, the approaches should be wide enough to permit fast movement of all vehicles anticipated for use on the facility, including emergency vehicles and mobile cranes.

2.7.3 WALKWAY WIDTH. Separate walkway structures should have a minimum clear width of 3 ft. Where the walkway is attached to a vehicle traffic lane, provide a minimum width of 2 ft 6 in clear, from curb to safety railing.

2.7.4 ROADWAY DECK ELEVATION. The requirements for pier and wharf deck elevation are also applicable to the approaches. Where the adjacent land is higher or lower than pier or wharf, the approach can be sloped up or down to serve as a transition ramp. For approaches longer than 100 ft (30.5m), the slope should be limited to 6 percent. For shorter approaches, the maximum slope should be 8 percent, 12H:1V.

2.7.5 NUMBER OF APPROACHES.

2.7.5.1 ONE APPROACH. For fueling and magnetic treatment/electromagnetic roll facilities, at least one single-lane approach structure should be provided, unless the facility is built as an island wharf or pier with access by watercraft.

2.7.5.2 TWO APPROACHES.

- Where volumes of vehicular movements are large, provide at least two approaches to ensure continuous uninterrupted traffic flows from pier or wharf to shore. At multiple-berth facilities, consider approach structures at least every 500 ft.
- Where the width of the pier or wharf is not sufficient to permit turning of vehicles, provide two approaches. Thus, vehicular traffic may enter and leave the facility without having to turn around. Since it is easier for a truck to negotiate a left turn, design traffic patterns to favor left turns.

2.7.5.3 RAILROAD ACCESS. Where rail access is planned for either crane or railroad, a separate approach is not necessary. However, consider a separate walkway. Consider approach slope limit for crane or railroad.

2.7.6 TURNING ROOM. At the intersection of approach and piers and wharves, provide fillets or additional deck area at corners to allow for ease in executing turns. Where a one-lane approach roadway is provided as the only access, provide the pier or wharf with sufficient turnaround space on the facility so that outgoing vehicles do not have to back out along the approach.

2.7.7 SAFETY BARRIERS. On all approaches, provide safety barriers adequate for the type of traffic using the facility (pedestrian, vehicular, and/or rail). However, safety barriers should not be provided in areas where mission operations, such as ship or small craft berthing, are performed. Rail only approaches do not normally require safety barriers. Provide traffic barriers between pedestrian and traffic lanes. Traffic and pedestrian barrier design must conform to AASHTO Guide Specifications for Bridge Railings and AASHTO Bridge Guide and Manual Interim Specifications.

2.8 STRUCTURAL TYPES.

2.8.1 GENERAL. The three major structural types for piers and wharves are open, solid, and floating. Open type piers and wharves are pile supported platform structures that allow water to flow underneath. Figure 2-7 illustrates the open type. Solid type uses a retaining structure such as anchored sheet pile walls or quaywalls, behind which a fill is placed to form the working surface. Solid type will prevent stream flow underneath. Figure 2-7 illustrates the solid structural type. Floating type is a pontoon structure that is anchored to the seabed through spud piles or mooring lines and connected to the shore by bridges or ramps. A floating double deck pier is shown in Figure 2-8.

2.8.2 SELECTION OF TYPE. Numerous factors influence the selection of one structural type over the other. Evaluate each of these factors against the construction and operating costs of the facility before a final decision is made on the structural type. Place greater emphasis on selecting the type that will withstand: unexpected berthing/mooring forces, adverse meteorological and hydrological conditions, and the corrosive effects of a harsh marine environment such that it will require little or no maintenance. The geotechnical characteristics of a given site, and economic analysis of alternate structural types will often dictate structural requirements. For instance, in areas with poor near surface soils but with good end bearing for piles, an open pile supported structure with a shallow bulkhead (or no bulkhead) will be most economical. Conversely, in areas with good near surface soils and poor pile bearing, a solid bulkhead may be more economical.

2.8.2.1 SHORELINE PRESERVATION. The structural type is seriously influenced by aquatic and plant life existing along the shore of the planned facility. In environmentally sensitive areas such as river estuaries, the solid-type wharf, which would disturb or destroy a considerable length of shoreline, should not be considered. Select the open structural type, which would have the least impact on the shoreline.

2.8.2.2 BULKHEAD LINE. When the facility extends beyond an established bulkhead line (the limit beyond which continuous solid-type construction is not permitted) use open type construction.

2.8.2.3 TIDAL OR STREAM PRISMS. Where it is required to minimize restrictions of a tidal or stream prism (the total amount of water flowing into a harbor or stream and out again during a tidal cycle) use open type construction.

2.8.2.4 LITTORAL DRIFT. Along shores where littoral currents transporting sand, gravel, and silt are present, use open type construction to mitigate shoreline erosion and accretion.

2.8.2.5 ICE. In general, open type structures are vulnerable and should be carefully investigated at sites where heavy accumulations of sheet or drift ice occur. Also, when ice thaws, large blocks of ice may slide down the piling, impacting adjacent batter or plumb piles. Thus, the solid type may be preferable at such sites.

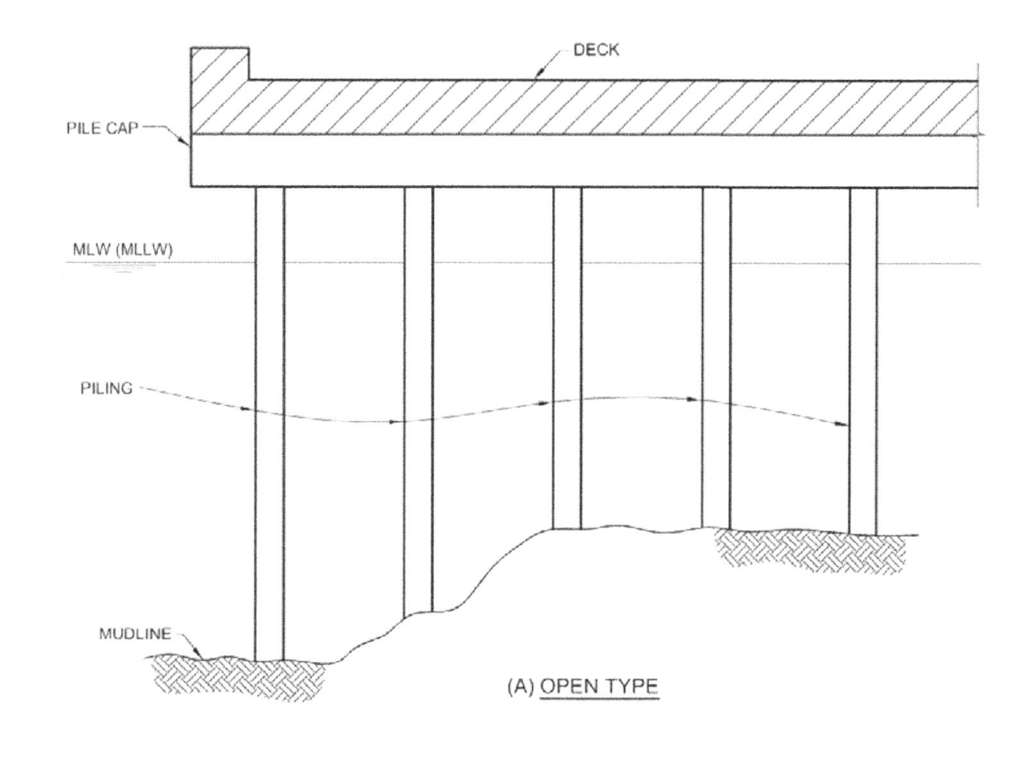

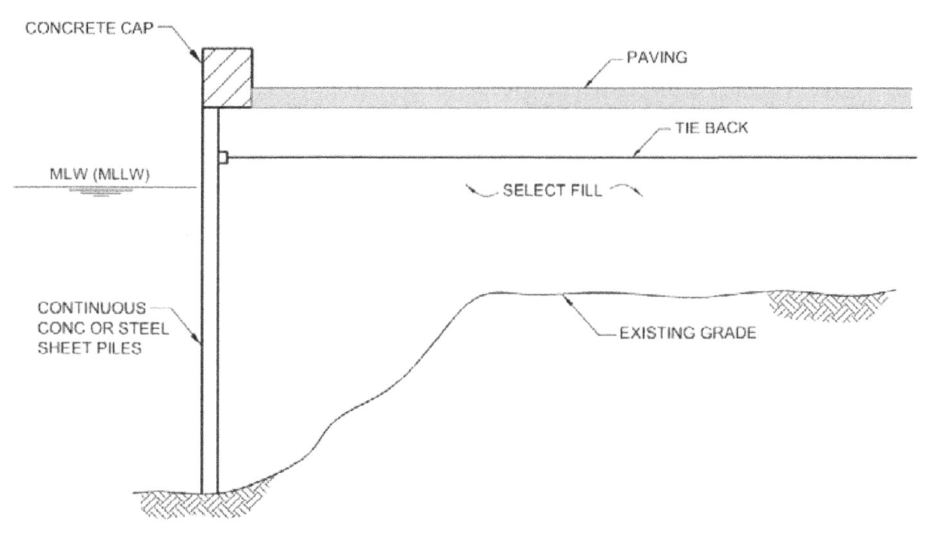

(B) SOLID TYPE

Figure 2-7

Open and closed types piers/wharves

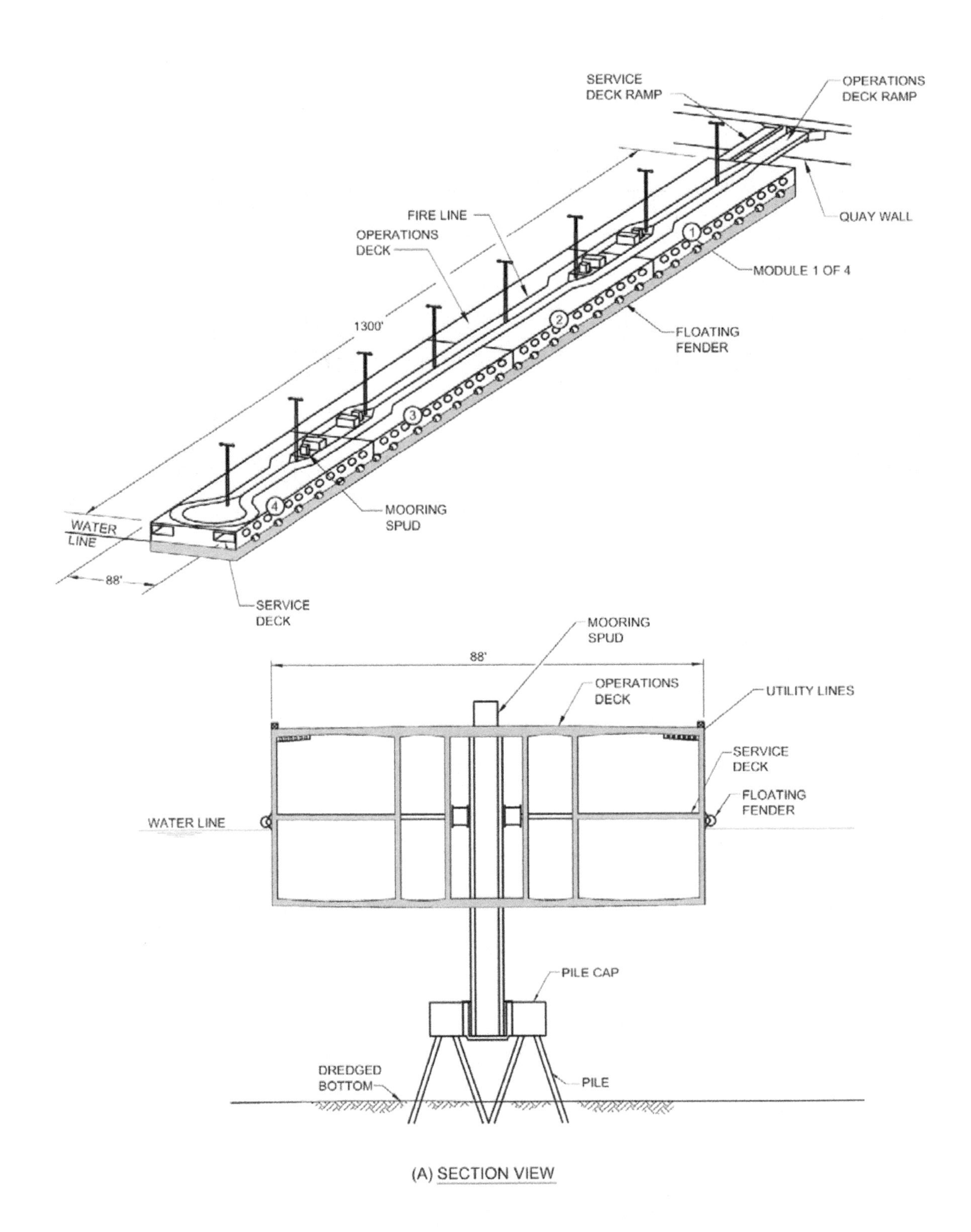

Figure 2-8

Floating double deck pier

2.8.2.6 EARTHQUAKE. In areas of high seismic activity, carefully consider construction utilizing sheet pile bulkheads or walls because of the high lateral earth pressures that can develop on the sheet piling. When a pile-supported platform (with curtain wall) is used for a wharf structure in conjunction with hydraulic fill, which is susceptible to liquefaction, consider a rock dike to resist the lateral forces that may be caused by liquefaction of the fill. The use of an engineered soil filter fabric between the rock dike and granular fill should also be considered. In areas of extremely high seismic risk, and where tsunamis and seiches are anticipated, seriously consider the floating type as it is less likely to be affected by or will suffer only minor damage from the seismic activity.

2.8.2.7 WATER DEPTHS. Consider open type construction in all depths of water when accommodating vessels, cargo vessels, and tankers. Depth limits for solid-type construction, utilizing sheet pile bulkheads, are imposed by the magnitude of the applied surcharge, subsurface conditions, and freeboard of the bulkhead above the low waterline. Generally, anchored sheet pile bulkheads may be considered in water depths up to 30 to 35 ft where favorable soil conditions exist. When greater water depths are required at solid type bulkhead structures, consider the use of relieving platforms, bulkheads consisting of reinforced high-strength steel sheet piles, or cellular construction.

2.8.2.8 SUBSURFACE CONDITIONS. Generally, subsurface conditions do not limit the use of open type construction. For almost all subsurface conditions, with the possible exception of rock close to the harbor bottom surface, suitable piles or caissons can be designed. Where rock is close to the surface and pile seating may be difficult and costly, consider cellular construction. When open type construction must be used in an area where rock is close to the surface, piles should be socketed and anchored into the rock. Consider sheet piling, used for bulkheads or retaining walls in conjunction with platform wharf structures or combination piers, only when subsurface conditions indicate that suitable anchorage and restraint for the toe of the sheet piling can be achieved and where select material is available for backfill.

2.8.2.9 FILL LOSS. When precast concrete and steel sheet pile bulkheads are used in pier and wharf construction, take special care to prevent fill leaching through the interlocks, causing subsidence of retained fill. Install a filter blanket or other method that could prevent or control fill leaching to reduce subsidence and consequent paving maintenance.

2.8.2.10 REMOTE AREAS. Facilities in remote areas may include modular, floating and elevated piers, wharves, and docks; powered and non-powered causeways; roll-on/roll-off floating platforms; and fabrication of special craft and bulk liquid transfer systems. Some examples are noted below:

- Steel Pontoon Wharf. Modular steel pontoon structures may be used for temporary facilities to berth ships up to loaded drafts of 30 ft. This structure type may be provided where it is not advisable to construct a fixed facility and at advanced bases where versatility and ease of deployment are required. The allowable uniform loading is limited, as is the capacity for mobile cranes.

- Pontoon Structures. Lighterage structures, when joined together, form a roadway to permit movement of vehicles, personnel and supplies from ship to shore during amphibious operations. In addition to bridging the gap between ships and the beach, floating causeways may be used as lighterage barges to transport vehicles and supplies to a wharf or to a beach, or may serve as piers for unloading small craft. NL pontoons are reinforced, welded steel cubes, 5 ft long, 7 ft wide, and 5 ft high, capable of accommodating HS-20 truck loading as specified by AASHTO. Pontoons are assembled into strings which are joined to form pontoon barges and pontoon bridge units. Thus, a 3 x 15 pontoon consists of 3 strings of 15 pontoons each. The primary NL lighterage structures are 3 x 15 causeway sections of both powered and non-powered variety. Two powered configurations include the Side Loadable Warping Tug (SLWT) and the Causeway Section Powered (CSP). Three non-powered configurations include: the Causeway Section Non-powered Intermediate (CSNP-IN), the Causeway

Section Non-powered Beach (CSNP-BE), and the Causeway Section Non-powered Offshore (CSNP-OS). These powered and non-powered causeway sections can work together in various configurations to form barge ferries, floating piers, elevated piers, and Roll-On/Roll-Off Discharge Facilities (RRDF) used to assist downloading RO/RO ships while moored in stream..

- Jack-up Barge. This type consists of a structural steel seaworthy barge provided with openings for steel caissons, which are lowered to the harbor bottom when the barge has been floated into final position. The barge may be completely outfitted during construction with ship fenders, deck fittings, and utilities including power, lighting, communications, water supply, sanitary facilities, etc., so that once it is jacked into position and utility tie-ins are made, it is ready to receive ships. Circular pneumatic gripping jacks, mounted on the deck above the caisson openings, permit the barge to be elevated in steps. The barge is loaded with steel caissons, a crane for pile erection, and other tools and materials required for the fieldwork, and is towed to the site. At the site, the barge is moved into approximate position and the caissons are dropped through the jacks and hull by the crane. The caissons, suspended above the harbor bottom and supported by engaging the jacks, are seated into the harbor bottom by dead weight. The barge-like deck is jacked to the required elevation and locked. Each caisson is then released from its jack and driven to refusal or required penetration. When all caissons are driven, the hull of the barge is welded to the caisson, the jacks are removed, and the caissons are cut off flush with the deck and capped with steel plates. In some situations, the caissons are filled with sand to avoid buoyancy problems. Jack-up barge type structures are also constructed using hydraulic jacks and open-trussed towers instead of pneumatic jacks and circular caissons.

- Template. This type involves the fabrication of the various structural components of the pier, transportation of the prefabricated units to the construction site by barge, and erection of the prefabricated units to form the completed facility. As noted under jack-up barge, the template type pier may be outfitted, beforehand, with the utilities, deck fittings, and services that are needed to produce a fully working berthing facility. The prefabricated structural steel units consist of templates, deck assemblies made up of cap beams or trusses and stringers, tubular piles, fender units, decking (timber or concrete), fittings, and miscellaneous hardware. The template is an assembly consisting of four or more tubular columns connected with tubular bracing and welded together to form a structure of height approximately equal to the depth of water in which it is to be installed. A floating crane is used to transfer the template from the transporting barge and position it on the harbor bottom. Steel piles are placed through the template tubular columns and driven to refusal or the required penetration. If the harbor bottom is very soft, the template is held in a suspended condition while the steel tubular piles are placed and driven through the template columns. After pile driving, the space between the piles and the template columns is filled with grout. As succeeding templates are erected, deck units, decking, fender units, and fittings are placed to form the completed marine facility. Based on past experiences, it is estimated that a prefabricated template type structure, 90 ft (27.4 m) wide x 600 ft (182.9 m) long, could be erected in about 21 days and a structure of the jackup type could be erected in about 3 days. An advantage of the jackup barge structure is that it can be moved and reused at other sites. For permanent facilities in remote areas, the floating type has advantages as the onsite construction is minimized.

-
- Container-Sized Modular Pontoons. Commercially available modular causeway systems are becoming increasingly popular for naval operations. The pontoons are made from container-sized modules that can be stored and transported in commercial container ships. The modules are lifted out and assembled on calm water or launched from a ship's deck. Cleats, bitts, and other hardware can be installed and removed as needed. With minor modifications, the individual modules can be structured to accommodate power units and fuel tanks for use as a powered causeway. The connector pieces are interchangeable and removable for rapid repairs. The modular causeway system makes it possible to

transport the causeway to the amphibious operations area on a container ship or an auxiliary crane ship for rapid deployment."

- The Modular Elevated Causeway (ELCAS(M)) is a temporary pierhead and 24 ft wide roadway made from connected ISO-compatible P40-Series steel pontoons elevated on piles and extending seaward across the surf zone to a distance up to 3,000 ft from the beach. The entire ELCAS(M) pier facility, including the necessary construction and installation equipment, is designed to be transported on one Transport-Auxiliary Crane Ship (TACS). The actual system installation length is determined by the requirement to reach a maximum water depth at the pierhead of 20 ft (6.1 m) Mean High Water; with the underside of the pier structure a minimum of 15 ft (4.6 m) above the Mean Low Water level based on a tidal range of 8 ft (2.4 m). The primary function of the system is to provide throughput of containerized cargo over the surfzone offloaded from lighterage at the pierhead, which is carried by trucks to the beach. Construction of the pier system is accomplished in a cantilever fashion from one constructed 40-ft (12.2 m) section to the next, from the beach out to the pierhead over the surfzone with no construction equipment operating from platforms on the water.

- The Amphibious Bulk Liquid Transfer System (ABLTS) is a self-contained fuel and water system capable of deploying 10,000 feet of 6-inch fuel hose line and 4-in (100 mm) water hose line from a designated ship to an ABLTS Beach Interface Unit (BIU) ashore.

2.8.2.11 CONSTRUCTION TIME. Where an existing pier or wharf has to be replaced, the floating type has the advantage of minimizing the "downtime." Conventional construction may take too long where the loss of berths cannot be tolerated. The floating type in such situations may turn out to be the most expedient.

2.8.2.12 SHIP CONTACT. In certain situations, where tugboats or camels are not available, sheet pile bulkheads located along the offshore face of pier and wharf structures may be less desirable than open type construction because of the greater danger for contact between the sheet piling and the bulbous bow or sonar dome of a ship during berthing and unberthing maneuvers.

2.8.2.13 TRACK MOUNTED CRANE. Where a track-mounted crane is required for the pier or wharf, the solid type may not be suitable. The susceptibility of the solid filled type to settlement and movement will make it very difficult to maintain the close tolerance required for rail gage, elevation, and alignment. The surcharge loading on the sheet pile will also be considerable. For such cases, use an independent pile supported track.

2.8.3 CONSTRUCTION. Several aspects of construction that are unique to each structural type should be considered.

2.8.3.1 OPEN. For open type wharves and landside ends of open piers, the following schemes should be considered for retaining upland fill:

- Platform on Piles and a Curtain Wall at the Inshore Face. The underwater slope should be as steep as possible, as limited by both constructional and geotechnical parameters, thus making the pile-supported platform narrow and more economical. In seismically active areas, where hydraulic fill susceptible to liquefaction is used for upland fill, a rock dike may be used instead of the granular fill dike to resist the lateral forces caused by liquefaction of the fill. The use of a filter fabric also should be considered at the hydraulic fill interface.
- Platform on Pile and a Sheet Pile Bulkhead at the Inshore Face. The sheet pile bulkhead permits a narrower platform. The cost tradeoff between platform width and bulkhead height should be investigated as the bulkhead may be found to cost as much or more than the pile supported platform width saved.

2.8.3.2 SOLID. Retaining structures may be constructed by the following means:

- Sheet Pile Bulkhead. The bulkhead consists of a flexible wall formed of steel or concrete sheet piling with interlocking tongue and groove joints and a cap of steel or concrete construction. The bulkhead is restrained from outward movement by placing an anchorage system above the low water level. Many types of anchorage systems can be used. The most common types in use in the United States consist of anchor rods and deadman anchors. The latter could be made of concrete blocks, steel sheet piling, or A-frames of steel, concrete, or timber piles. In countries outside the United States, an anchorage system consisting of piles, attached near the top of the sheet pile bulkhead and extending at batters up to 1 on 1 to embedment in firm material, is often used. Rock or earth anchors consisting of high-strength steel rods or steel prestressing cables are sometimes preferred in place of the anchor batter piles. Provide granular free-draining material adjacent to sheet pile bulkheads, extending from dredged bottom to underside of pavement on grade. Grade this material to act somewhat as a filter to limit subsequent loss of fines through the sheet pile interlocks. Placement of free-draining material should be in stages, commencing at the intersection of sheet piling and dredged bottom and progressing inshore. Eliminate mud and organic silt pockets. In general, do not consider hydraulic fill for backfill unless provision is made for the effects of fill settlement, potential liquefaction of fill in seismic zones, and high pressure exerted on sheet piling. Consider vibro-compaction for consolidation of hydraulic fill. In areas with tidal ranges greater than 4 ft (1.2 m,) provide 2 in (51 mm) diameter weep holes for the sheet piles above the mean low water level. When weep holes are used, provide graded filters to prevent loss of finer backfill material. Provide openings in pavement or deck for replenishment of material in order to compensate for loss and settlement of fill. In general, flexible pavement using asphaltic concrete is preferred over rigid pavement with portland cement concrete, as it is more economical to maintain and better able to accommodate underlying settlement.

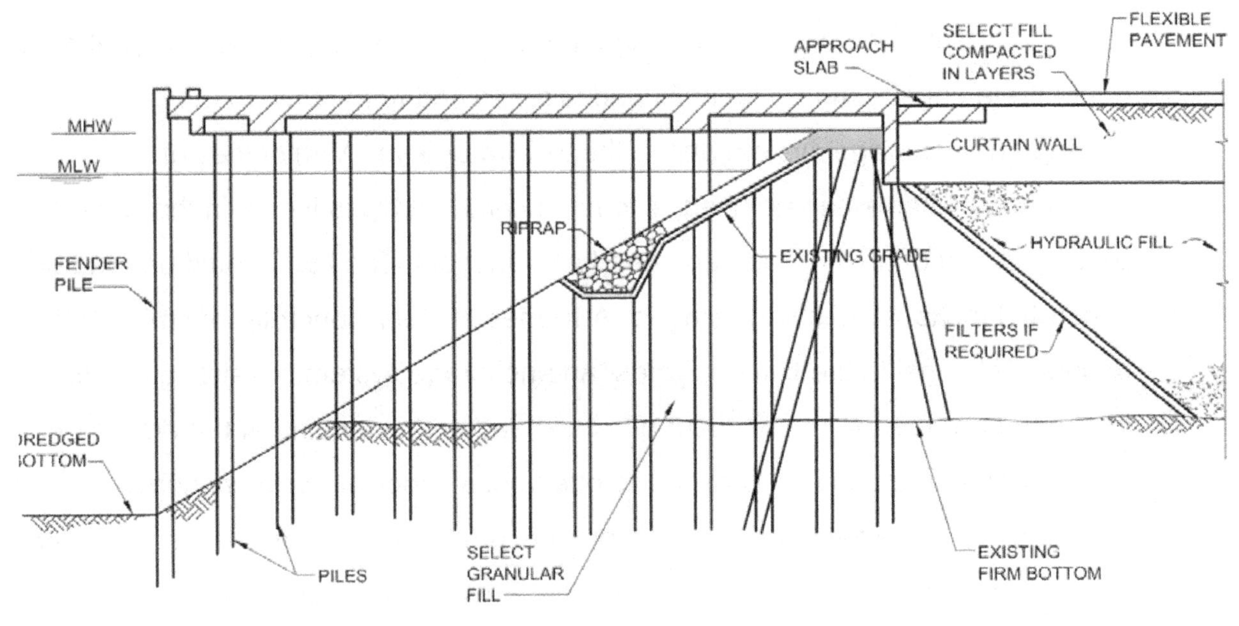

(A) WITH CURTAIN WALL

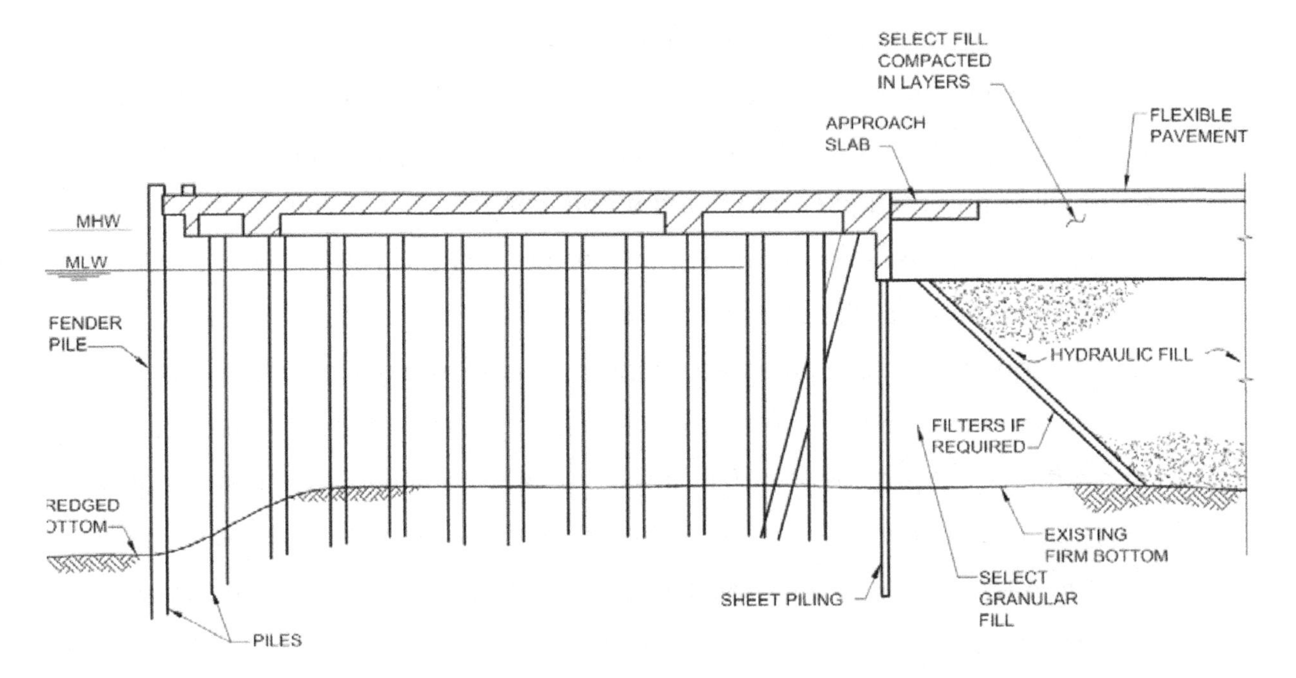

(B) WITH SHEET PILE BULKHEAD

Figure 2-9

Open type wharf concepts

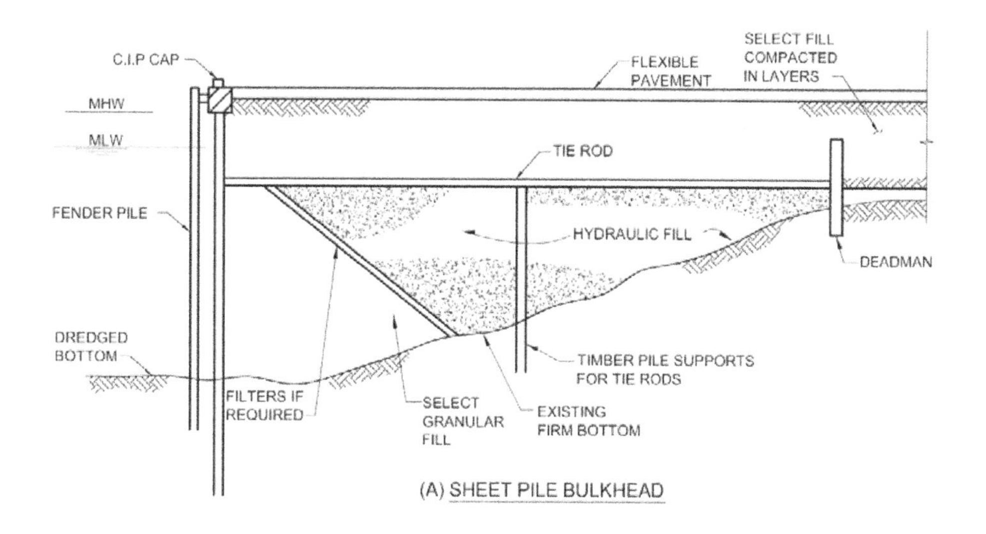

(A) SHEET PILE BULKHEAD

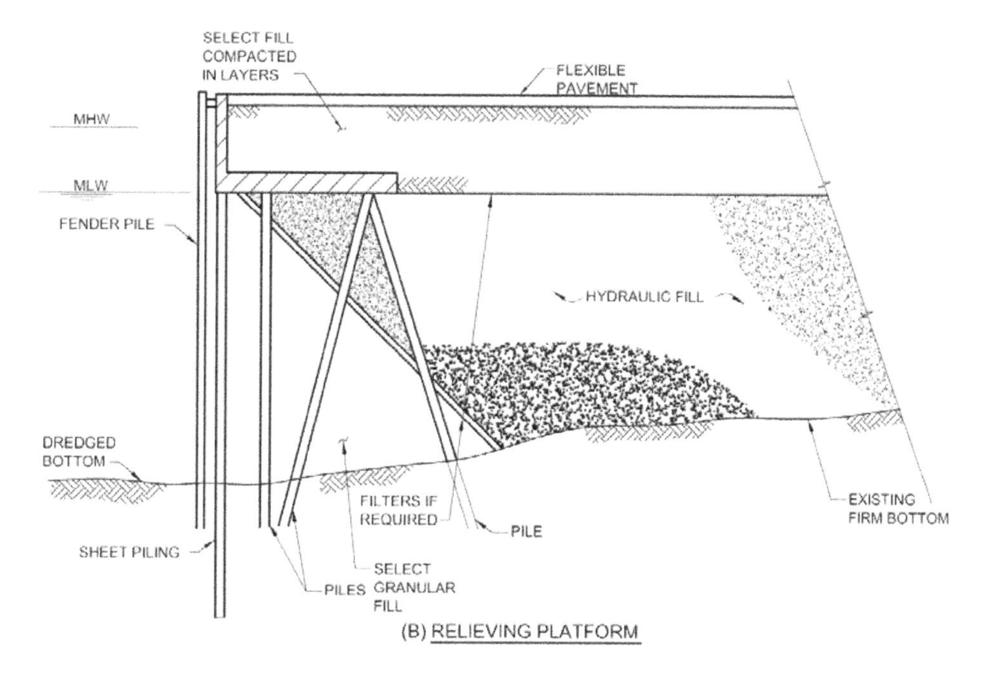

(B) RELIEVING PLATFORM

Figure 2-10

Closed type wharf concepts

used in conjunction with a sheet pile bulkhead to reduce the lateral load on the sheet piling created by heavy surcharges and earth pressures. See Figure 2-10. Lateral restraint is provided by the batter piles supporting the relieving platform. A variation of this type of construction is to use only vertical piles for the relieving platform and to furnish an independent anchorage system consisting of tie rods and deadman, similar to the types specified for sheet pile bulkheads.

- c. Cellular Construction Consisting of Sheet Pile Cells. For design procedures and selection of type, see UFC 3-220-01N, Geotechnical Engineering Procedures for Foundation Design of Buildings and Structures. Cellular structures are gravity retaining structures formed from the interconnection of straight steel sheet piles into cells. Strength of cellular structures derives from resistance to shear caused by friction of the tension in the sheet pile interlocks and also from the internal shearing resistance of the fill within the cells. Accordingly, clean granular fill materials such as sand and gravel are usually used to fill the cells. Extreme care must be exercised in the construction of cellular structures because excessive driving onto boulders or uneven bedrock may cause ruptured interlocks, which can later unzip under hoop tension (from filling) and cause failures of the cell. Movement and expansion of cells must be compensated for during construction of the cells and fill placement carefully controlled if satisfactory alignment of the face of the wharf is to be maintained. Cellular structures are classified according to the configuration and arrangement of the cells. Three basic types are discussed below:

    - Circular. This type consists of individual large-diameter circles connected together by arcs of smaller diameter. Each cell may be completely filled before construction of the next cell is started. Construction of this type is easier than the diaphragm type because each cell is stable when filled and thus may be used as a platform for construction of adjacent cells. Because the individual cells are self-supporting units, accidental loss of one cell will not necessarily endanger adjoining cells. Compared to a diaphragm type cellular structure of equal design, fewer piles per linear foot of structure are required. The diameter of circular cells is limited by the maximum allowable stresses in the sheet pile interlocks and, when stresses are exceeded, cloverleaf cells are used.

- ○ • Diaphragm. This type consists of two series of circular arcs connected together by diaphragms perpendicular to the axis of the cellular structure. The width of cells may be widened by increasing the length of the diaphragms without raising interlock stress, which is a function of the radius of the arc portion of the cell. Cells must be filled in stages so that the heights of fill in adjoining cells are maintained at equal levels to avoid distortion of the diaphragm walls. Diaphragm type cells present a flatter faced wall than circular cells and are considered more desirable for marine structures.

- ○ • Cloverleaf. This type is a modification of the circular cell type and is generally used in deep water where the diameter required for stability would result in excessively high interlock stress if diaphragms were not added.

- ○

- d. Reinforced Concrete Caisson. In this type of construction, concrete caissons are cast in the dry, launched, and floated to the construction site where they are sunk on a prepared foundation. The caisson is filled with gravel or rock and a cast-in-place retaining wall is placed from the top of the caisson to the finished grade. This type of construction is prevalent in countries outside the United States.

- e. Precast Concrete Blocks. This form of solid wharf is a gravity type wall made up of large precast concrete blocks resting on a prepared bed on the harbor bottom. A select fill of granular material is usually placed in the back of the wall to reduce lateral earth pressures. This type of construction is popular outside the United States.

2.8.3.3 FLOATING. Construction of the floating type usually requires a flood basin, graving dock, or drydock. The units are essentially constructed in the dry and floated out and transported (on their own or on barges) to the site. Availability of such a facility and transportation of the floating units through open ocean waters and restricted inland waters for deployment at the site are serious considerations. In this respect, the floating type has a significant advantage over others in that the bulk of construction activity can be shifted to other parts of the country where labor, economic, and environmental conditions are more favorable.

2.9 HYDRAULIC FILL. The soil drawn up by the suction head of a dredge, pumped with water through a pipe, and deposited in an area being filled or reclaimed is referred to as "hydraulic fill." At port and terminal facilities, where land is not available onshore and where dredging is required to provide adequate water depths for vessels at berths and approach channels, hydraulic fill is commonly used for land reclamation because of its availability and low cost. Hydraulic fill may be of good quality, consisting of granular materials, or may consist of plastic organic silt, which is considered poor quality. When hydraulic fill is used, the stability of the structure retaining the fill must be investigated, taking into consideration the effects of adjacent surcharge loadings in addition to the loadings from the fill. The placement of a select granular fill adjacent to the retaining structure may be required if the hydraulic fill is of poor quality. Hydraulic fill is in a loose condition when placed. To avoid fill settlements due to loadings from other structures, stacked cargoes, and mobile equipment, stabilization of the fill may be required. In areas of seismic activity, investigate the liquefaction of hydraulic fills. Stability with regard to both settlements and liquefaction may be enhanced by methods such as deep densification or by use of sand drains. Material other than hydraulic fill should be used when the cost of material obtained from onshore borrow areas is cheaper than the cost of material obtained from offshore borrow areas or where good quality fill material is required and is not available offshore.

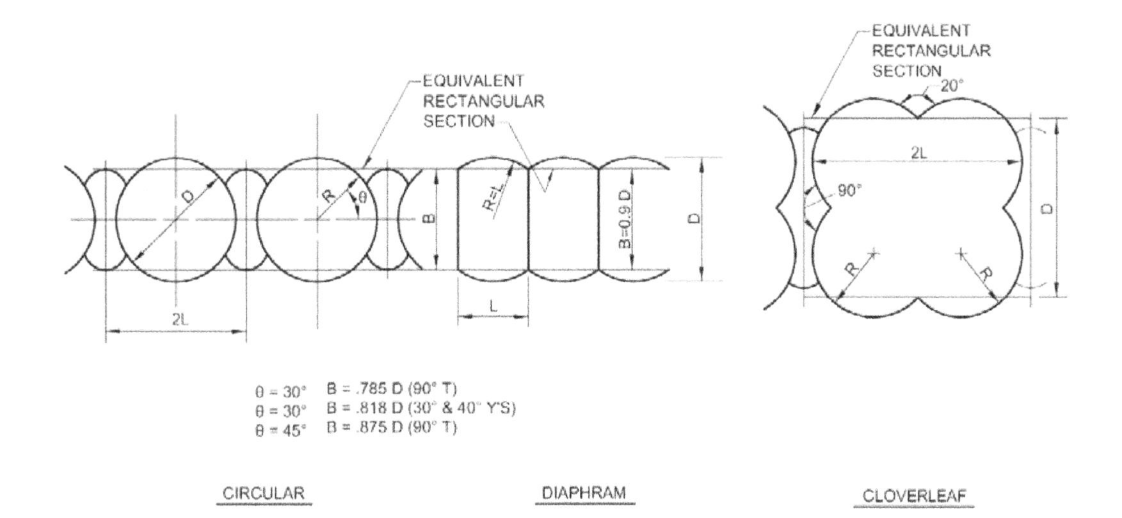

(A) TYPES OF CELLULAR CONSTRUCTION

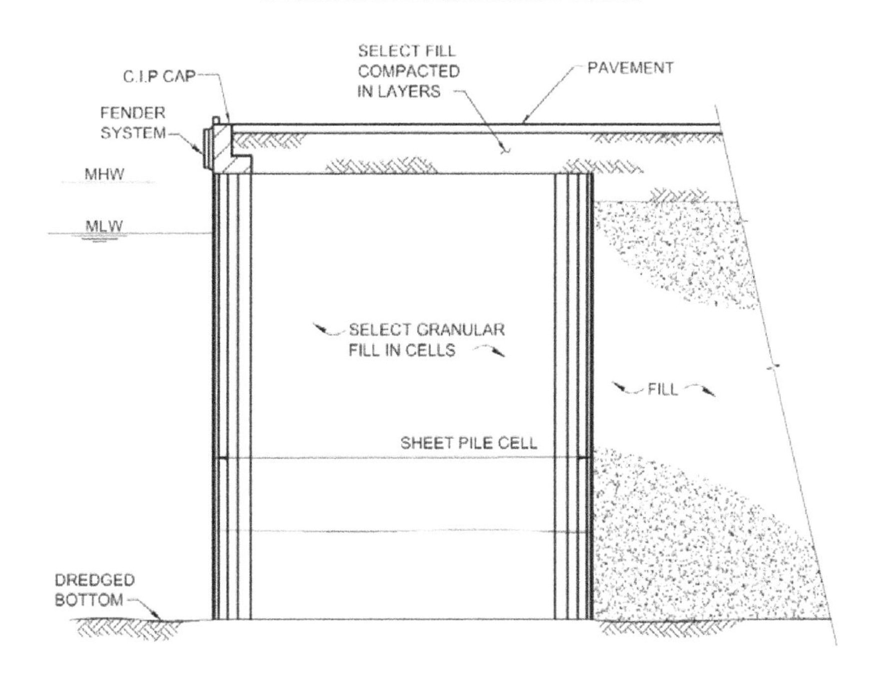

(B) TYPICAL SECTION

Figure 2-11

Solid type cellular construction

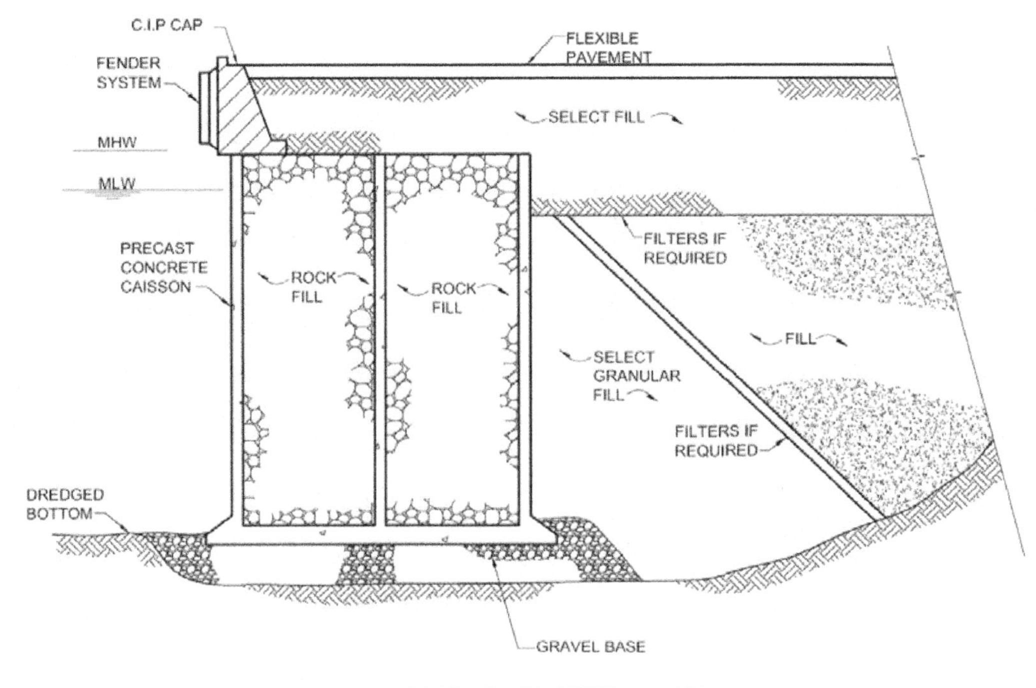

(A) PRECAST CONCRETE CAISSON

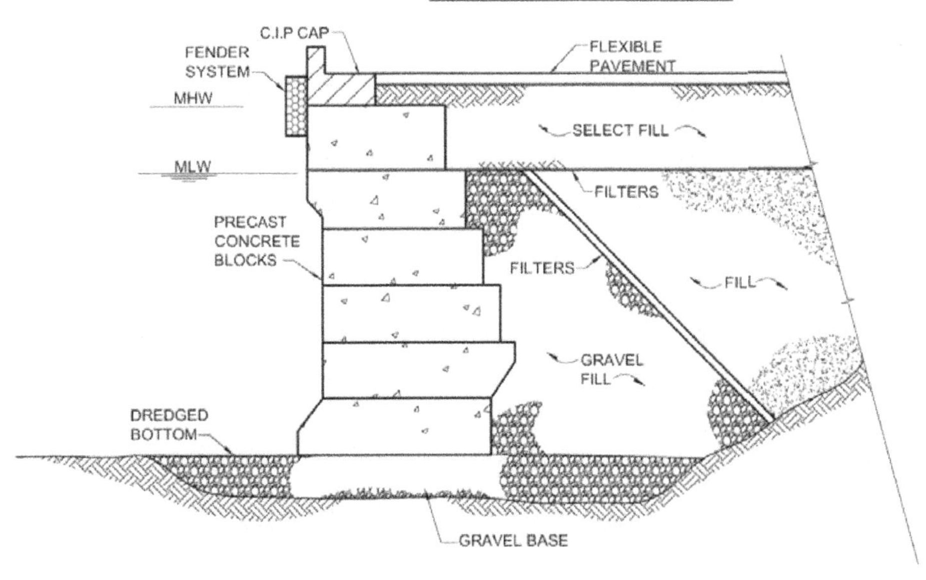

(B) PRECAST CONCRETE BLOCKS

Figure 2-12

Solid type, caisson and concrete block construction

CHAPTER 2
DESIGN LOADS

1. GENERAL. Where loading conditions exist that are not specifically identified in this discussion, rely on accepted industry standards. However, in no case will other standards supersede the requirements provided by this publication.

2. DEAD LOADS.

2.1 GENERAL.

The dead load consists of the weight of the entire structure, including all the permanent attachments such as mooring hardware, light poles, utility booms, brows, platforms, vaults, sheds, and service utility lines. A realistic assessment of all present and future attachments should be made and included. Overestimation of dead loads generally will not adversely affect the cost of the structure. However, overestimation of dead loads would not be conservative for tension or uplift controlled design. Also, for floating piers and wharves, overestimating of dead loads would overstate draft and could have a significant effect on cost.

2.2 UNIT WEIGHTS.

Use actual and available construction material weights for design. See Table 3-1 for unit weights that should be used for construction materials (unless lesser unit weights can be demonstrated by local experience):

Material	Unit Weight, pcf
Steel or cast steel	490 pcf
Cast iron	450 pcf
Aluminum alloys	175 pcf
Timber (untreated)	40 to 50 pcf
Timber (treated)	45 to 60 pcf
Concrete, reinforced (normal weight)	145 to 160 pcf
Concrete, reinforced (lightweight)	90 to 120 pcf
Compacted sand, earth, gravel, or ballast	150 pcf
Asphalt paving	135 to 150 pcf

Table 3-1

Unit weights

3. VERTICAL LIVE LOADS.

3.1 GENERAL. Although a number of loading conditions will be presented in subsequent sections, the advent and subsequent heavy usage of mobile cranes on piers will generally produce the controlling loading condition. For example, on a general purpose berthing pier, a design drawing will typically present different design live loads including uniform, vehicular (i.e. HS 20-44), forklift, and maximum outrigger float load. The outrigger load from the mobile crane invariably controls with the other loads more or less provided for informational purposes.

3.2 UNIFORM LOADING. See Table 3-2 for recommended uniform loadings for piers and wharves. Impact is not applied when designing for uniform loads.

Classification	Uniform Loading (psf)	Mobile Crane (tons)	Rail-Mounted Crane (tons)	Other Handling Equipment (tons)
Ammunition	600	90	--	20-lift truck
Berthing (carriers)	800	140	--	20-lift truck
Berthing (all others)	600	90	--	20-lift truck
Berthing (submarines)	600	100	--	20-lift truck
Fitting-out	800	140	60/151 Portal	20-lift truck
Repair	600	140	60/151 Portal	20-lift truck
Fueling	300	50	--	10-lift truck
Supply (general cargo)	750	140	--	20-lift truck
Supply (containers)	1,000	140	40/50 (long tons) Container	20-lift truck 33-straddle carrier

*psf can be converted to pascals (Pa) by multiplying by 47.88026.

Table 3-2

Vertical live loads for pier and wharf decks

3.3 TRUCK LOADING. Calculate truck wheel loads in accordance with the American Association of State Highway and Transportation Officials (AASHTO) Standard Specifications for Highway Bridges except as modified below. In the design of slabs, beams, and pile caps, apply an impact factor of 15 percent. Structural elements below the pile caps need not be designed for impact. When truck loading is transferred through 1.5 ft (0.45 m) or more of crushed rock ballast and paving, and for filled construction, the impact forces need not be considered for design. Also, check with local activity for use of an overload vehicle such as weapons cradles, missile hauling vehicles, etc. which may have significantly higher wheel loads.

3.4 RAIL-MOUNTED CRANE LOADING

3.4.1 PORTAL CRANES. For design of new piers and wharves, the specific wheel load information about the crane(s) to be used should be obtained. If the crane resources are not known at the time of design, consult the appropriate authority for design loading and crane procurement. When choosing design wheel loads, consideration should be given to flexibility allowing for different makes of cranes and ultimate crane replacement as well as future use of the pier or wharf. Piers and wharves have greater longevity than rail-mounted cranes. To this end, a wheel load of 110,000 lbs (4990 Kg) minimum on 4-ft (1.2m) centers allows for a practical range of options. See Table 3-2 for typical rated capacities of cranes for a variety of pier and wharf deck uses. Figure 3-1 lists a sampling of wheel loads for 60 ton and 151 ton capacity portal cranes. These cranes were procured around capacities of existing piers and wharves and have wheel loads that are somewhat restrictive for new design. The values provided are typical for existing equipment used and are useful for design feasibility studies on existing structures. The rail gage should be approximately 30 feet minimum for a 60 ton capacity portal crane, and up to 40 feet for a portal crane of 100 tons capacity greater. For handling of fuel containers at repair or fitting out piers, portal cranes with up to 151-ton capacity are required.

3.4.2 CONTAINER CRANES. See Figure 3-2 for crane configuration and wheel loads of container cranes and Table 3-2 for the rated capacities of container cranes applicable to piers and wharves. The configuration and wheel loads are derived from several manufacturers and should be used only as a guide. A recent trend in the shipment of containerized cargo is to use larger ships, and this is the driving force in the design of container cranes. The size of the ship to be serviced will dictate the capacity, configuration, operational characteristics, and gage of the crane. The evolution in container crane design has been to increase the gage to 100 ft (30.5 m) and outreach on the boom to 150 ft (45.7 m) and larger while maintaining the lift capacity between 40 and 50 long tons. Hence, specific information on the size of the ship to be serviced and details from the crane manufacturer needs to be obtained for final crane design.

3.4.3 WHEEL LOAD UNCERTAINTY. Portal and container cranes are usually procured separately from the construction. The maximum allowable wheel loads are normally specified on the crane procurement documents. The number and spacing of wheels are critical to the structural capacity of an existing facility and structural design of a new facility. Having established the required capacity and configuration of a crane, the designer of a pier or wharf should consult with the appropriate authority and obtain wheel loads for which the supporting structure should be designed. In the absence of hard information, the 110,000 lb (4990 kg) wheel load presented in paragraph 3.4.1 may be used for portal cranes. However, the container crane wheel loads presented in

Figure 3-2 are only provided as a guide. The design characteristics noted in paragraph 3.4.2 must be determined in order to determine realistic wheel loads.

3.4.4 IMPACT. Apply an impact factor of 25 percent to the maximum listed wheel loads for the design of deck slab, crane girders, and pile caps. The impact factor is not applicable to the design of piles and other substructure elements.

3.5 MOBILE CRANE LOADINGS. The deck design for open and floating structural types of piers and wharves is usually controlled by mobile crane loading. However, the operational constraints imposed by under specifying mobile crane loadings are severe. Consequently, take care to specify realistic loading. Refer to Table 3-2 for designated mobile cranes applicable to each functional type of pier and wharf. As a minimum design the pier or wharf for the designated mobile crane, however, check with the local activity to confirm whether a crane larger than that designated could be used at the facility.

3.5.1 WHEEL LOADS. See Figure 3-3 for wheel loads for 50-, 70-, 90-, 115-, and 140-ton capacity mobile cranes. The information in Figure 3-3 is for typical truck cranes, although rough-terrain type mobile cranes are also used on piers and wharves. Tire contact area should be as defined by AASTHO. As a rule of thumb, ground pressures for "on rubber" lifts are about 10 percent higher than tire inflation pressure. Crane manufacturers recommend that the majority of lifts be made on outriggers. In addition, capacities for "on rubber" lifts are substantially less than for "on outrigger" lifts. Hence, loads for "on rubber" lifts are not listed. Design all piers and wharves and their approaches for the wheel loads from the designated truck crane.

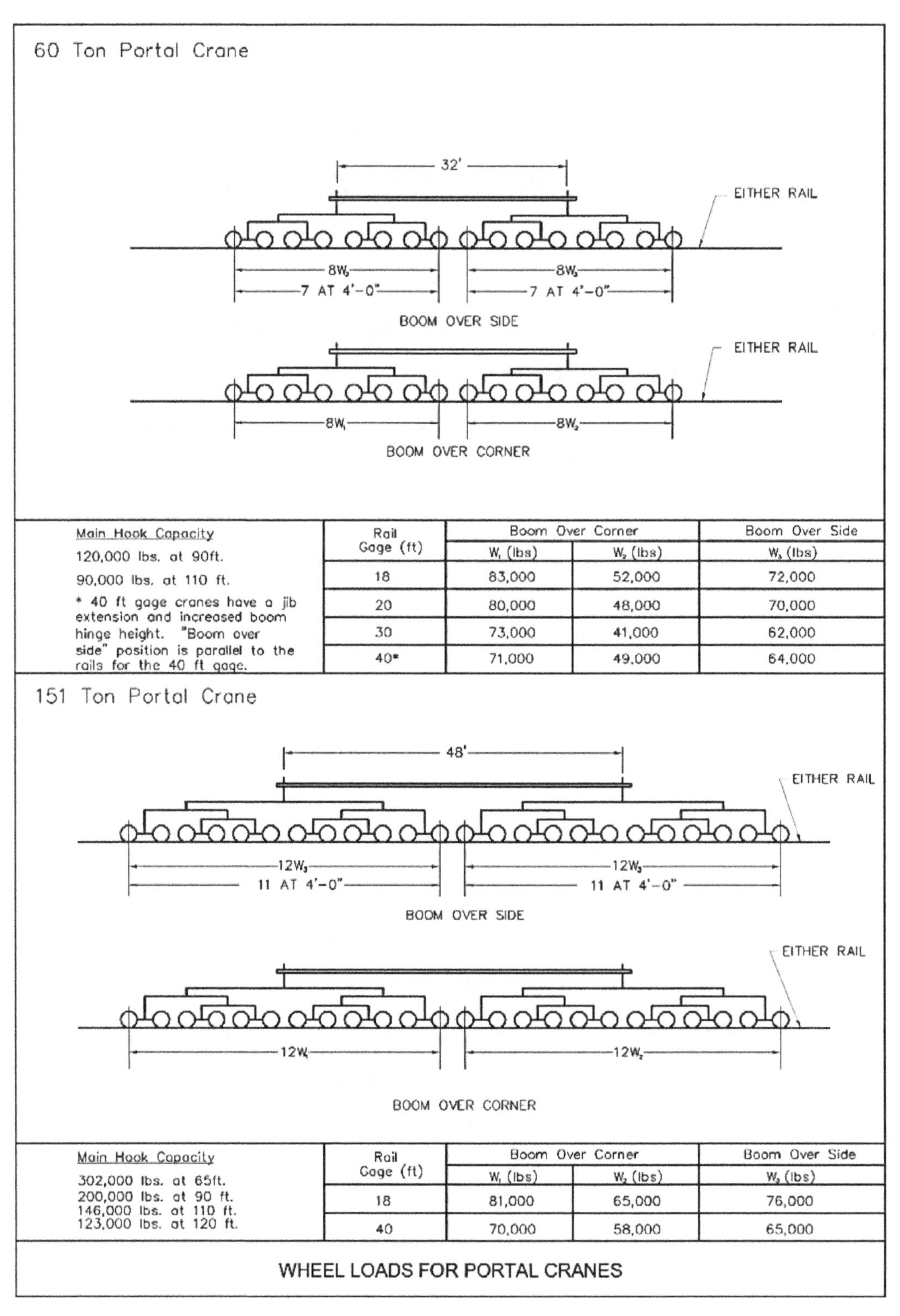

Main Hook Capacity	Rail Gage (ft)	Boom Over Corner		Boom Over Side
		W_1 (lbs)	W_2 (lbs)	W_3 (lbs)
120,000 lbs. at 90ft.	18	83,000	52,000	72,000
90,000 lbs. at 110 ft.	20	80,000	48,000	70,000
* 40 ft gage cranes have a jib extension and increased boom hinge height. "Boom over side" position is parallel to the rails for the 40 ft gage.	30	73,000	41,000	62,000
	40*	71,000	49,000	64,000

Main Hook Capacity	Rail Gage (ft)	Boom Over Corner		Boom Over Side
		W_1 (lbs)	W_2 (lbs)	W_3 (lbs)
302,000 lbs. at 65ft. 200,000 lbs. at 90 ft.	18	81,000	65,000	76,000
146,000 lbs. at 110 ft. 123,000 lbs. at 120 ft.	40	70,000	58,000	65,000

WHEEL LOADS FOR PORTAL CRANES

Figure 3-1

Wheel Loads for Portal Cranes

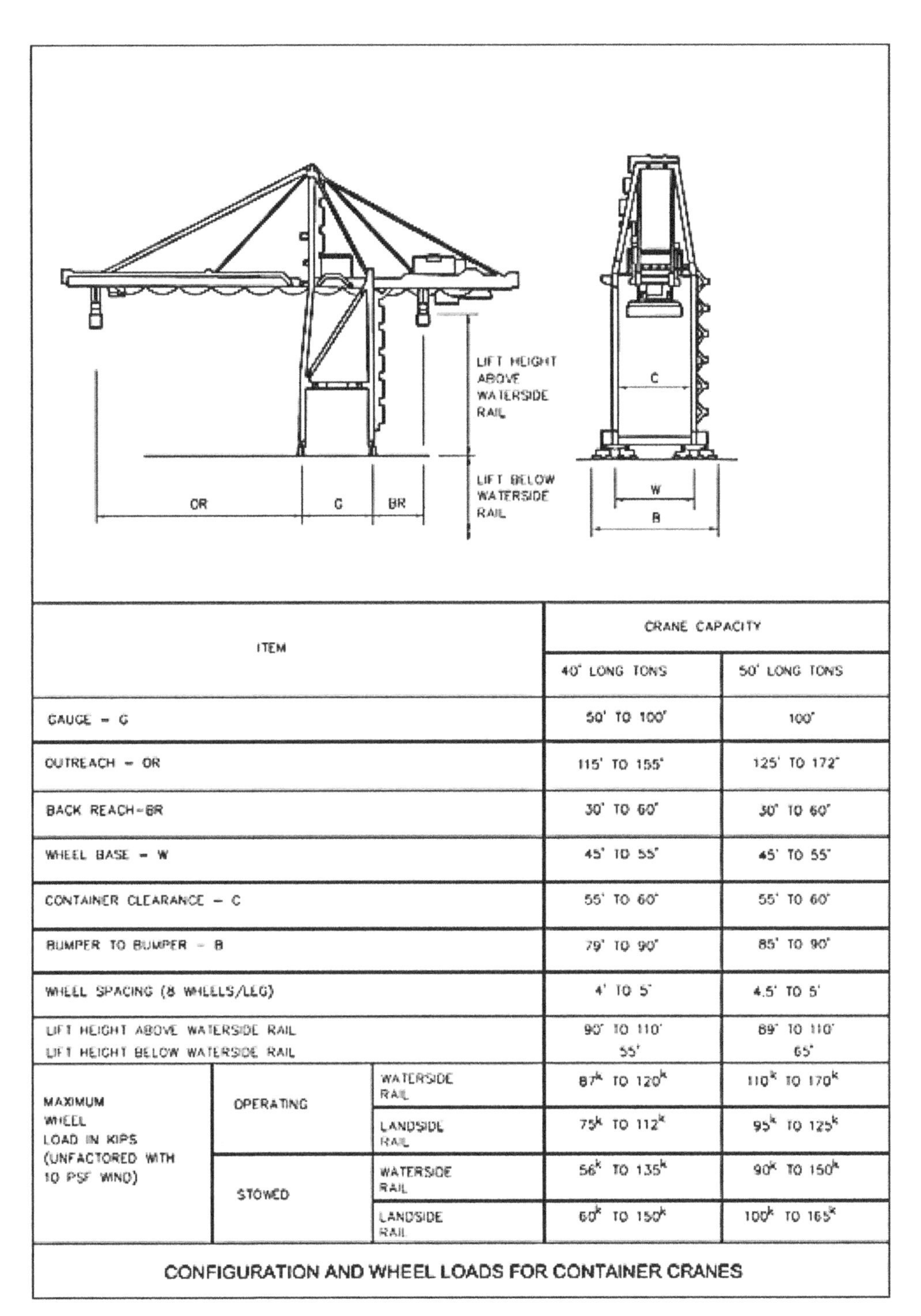

ITEM			CRANE CAPACITY 40' LONG TONS	CRANE CAPACITY 50' LONG TONS
GAUGE – G			50' TO 100'	100'
OUTREACH – OR			115' TO 155'	125' TO 172'
BACK REACH – BR			30' TO 60'	30' TO 60'
WHEEL BASE – W			45' TO 55'	45' TO 55'
CONTAINER CLEARANCE – C			55' TO 60'	55' TO 60'
BUMPER TO BUMPER – B			79' TO 90'	85' TO 90'
WHEEL SPACING (8 WHEELS/LEG)			4' TO 5'	4.5' TO 5'
LIFT HEIGHT ABOVE WATERSIDE RAIL LIFT HEIGHT BELOW WATERSIDE RAIL			90' TO 110' 55'	89' TO 110' 65'
MAXIMUM WHEEL LOAD IN KIPS (UNFACTORED WITH 10 PSF WIND)	OPERATING	WATERSIDE RAIL	87^k TO 120^k	110^k TO 170^k
		LANDSIDE RAIL	75^k TO 112^k	95^k TO 125^k
	STOWED	WATERSIDE RAIL	56^k TO 135^k	90^k TO 150^k
		LANDSIDE RAIL	60^k TO 150^k	100^k TO 165^k

CONFIGURATION AND WHEEL LOADS FOR CONTAINER CRANES

Figure 3-2

Configuration and wheel loads for container cranes

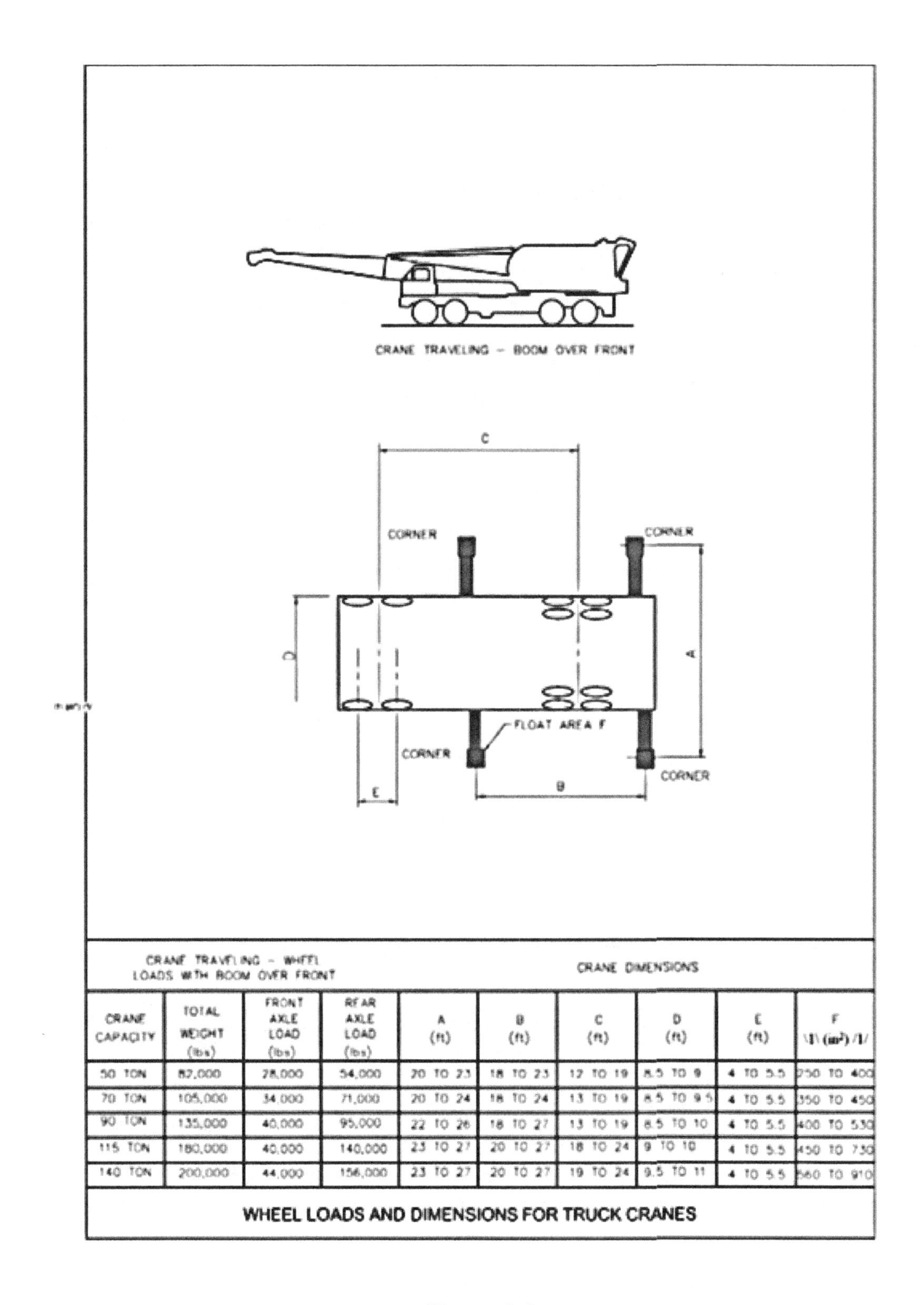

CRANE TRAVELING – WHEEL LOADS WITH BOOM OVER FRONT				CRANE DIMENSIONS					
CRANE CAPACITY	TOTAL WEIGHT (lbs)	FRONT AXLE LOAD (lbs)	REAR AXLE LOAD (lbs)	A (ft)	B (ft)	C (ft)	D (ft)	E (ft)	F (in^2)
50 TON	82,000	28,000	54,000	20 TO 23	18 TO 23	12 TO 19	8.5 TO 9	4 TO 5.5	250 TO 400
70 TON	105,000	34,000	71,000	20 TO 24	18 TO 24	13 TO 19	8.5 TO 9.5	4 TO 5.5	350 TO 450
90 TON	135,000	40,000	95,000	22 TO 26	18 TO 27	13 TO 19	8.5 TO 10	4 TO 5.5	400 TO 530
115 TON	180,000	40,000	140,000	23 TO 27	20 TO 27	18 TO 24	9 TO 10	4 TO 5.5	450 TO 730
140 TON	200,000	44,000	156,000	23 TO 27	20 TO 27	19 TO 24	9.5 TO 11	4 TO 5.5	560 TO 910

WHEEL LOADS AND DIMENSIONS FOR TRUCK CRANES

Figure 3-3

Wheel loads and dimensions for truck cranes

3.5.2 OUTRIGGER FLOAT LOADS. Table 3-3 lists outrigger float loads for different capacity cranes.

Capacity (tons)	Radius (ft)	Boom Over Corner (lbs)	Boom Over Back or Side (ea)(lbs)
50	20 and less 30 40 50 and more	97,000 83,000 78,000 72,000	89,000 69,000 60,000 55,000
70	20 and less 30 40 50 and more	130,000 115,000 104,000 99,000	114,000 90,000 85,000 78,000
90	20 and less 30 40 50 and more	160,000 144,000 130,000 117,000	144,000 120,000 107,000 98,000
115	20 and less 30 40 50 and more	195,000 180,000 156,000 144,000	183,000 151,000 131,000 120,000
140	20 and less 30 40 50 60 and more	232,000 222,000 211,000 198,000 187,000	207,000 176,000 164,000 150,000 142,000

Table 3-3

Outrigger Float Loads for Mobile Cranes

The maximum single float load from a boom over corner position and maximum concurrent pair of float loads from a boom over side and back positions are listed. Typically, the float loads are at the maximum when lifting the rated load at a short radius (20 ft (6.1 m) and less) and should be used for design. However, for existing piers and wharves, the other listed loads may be used to analyze deck capacity. Typical outrigger float spacing is shown in Figure 3-3. Apply outrigger float loads to a 1.5 ft by 1.5 ft (0.46 m by 0.46 m) area unless actual float size is known, in which case use the actual float size for analysis.

3.5.3 IMPACT. Apply an impact factor of 15 percent for all wheel loads when designing slab, beams, and pile caps. The impact factor need not be applied when designing piles and other substructure elements, when designing filled structures, and where wheel loads are distributed through paving and ballast (1.5 ft (0.46 m) or more).

3.6 FORKLIFT AND STRADDLE CARRIER LOADINGS.

3.6.1 FORKLIFTS. See Figure 3-4 for wheel loads from forklifts. Determine contact areas for wheel loads in accordance with AASHTO. For hard rubber wheels or other wheels not inflated, assume the wheel contact area to be a point load.

3.6.2 STRADDLE CARRIERS. See Figure 3-5 for wheel loads for a straddle carrier. The straddle carrier shown is capable of lifting a loaded 20-ft (6.1 m) container or a loaded 40-ft (12.2 m) container.

3.6.3 IMPACT. Apply an impact factor of 15 percent to the maximum wheel loads in the design of slabs, beams and pile caps. The impact factor need not be applied when designing piles and other substructure elements, when designing filled structures, and where wheel loads are distributed through paving and ballast (1.5 ft (0.46 m) or more).

3.7 LOADING ON RAILROAD TRACKS. For freight car wheel loads, use a live load of 8000 lbs/ft of track corresponding to Cooper E-80 designation of the American Railway Engineering Association (AREA) Manual for Railway Engineering. In the design of slabs, girders, and pile caps, apply an impact factor of 20 percent. Impact is not applicable for the design of piles and filled structures, or where loads are distributed through paving and ballast (1.5 ft (0.46 m) or more).

3.8 BUOYANCY. Typically, piers and wharf decks are not kept low enough to be subjected to buoyant forces. However, portions of the structure, such as utility trenches and vaults, may be low enough to be subject to buoyancy forces, which are essentially uplift forces applied at the rate of 64 psf (3064.3 Pa) of plan area for every foot of submergence below water level.

3.9 WAVE LOADING. For piers and wharves exposed to waves which may produce significant lateral or hydrostatic forces, determine wave loading in accordance with the procedures defined in the technical literature.

3.10 APPLICATION OF LOADINGS.

3.10.1 CONCENTRATED LOADS. Wheel loads and outrigger float loads from designated pneumatic-tired equipment, such as trucks, truck cranes, forklifts, and straddle carriers may be oriented in any direction and the orientation causing the maximum forces on the structural members should be considered for design. Significant loads from fuel containers at Repair or Fitting Out Piers exceeding 300 kips (1334.5 kN) may be encountered. Design trench covers, utility trench covers, and access hatch covers to handle the concentrated loads, where they are accessible to mobile equipment. However, designated areas on the pier deck may be exempted from wheel loads or outrigger float loads, or designed for lesser loads, when curbs, railings, and other physical barriers are provided to isolate those areas from vehicle access. Concentrated wheel loads from these vehicles are applied through small "footprints" to the deck structure. The distribution of these loads and computation of maximum moments and shears may be in accordance with the AASHTO Standard Specification for Highway Bridges. However, this method is conservative.

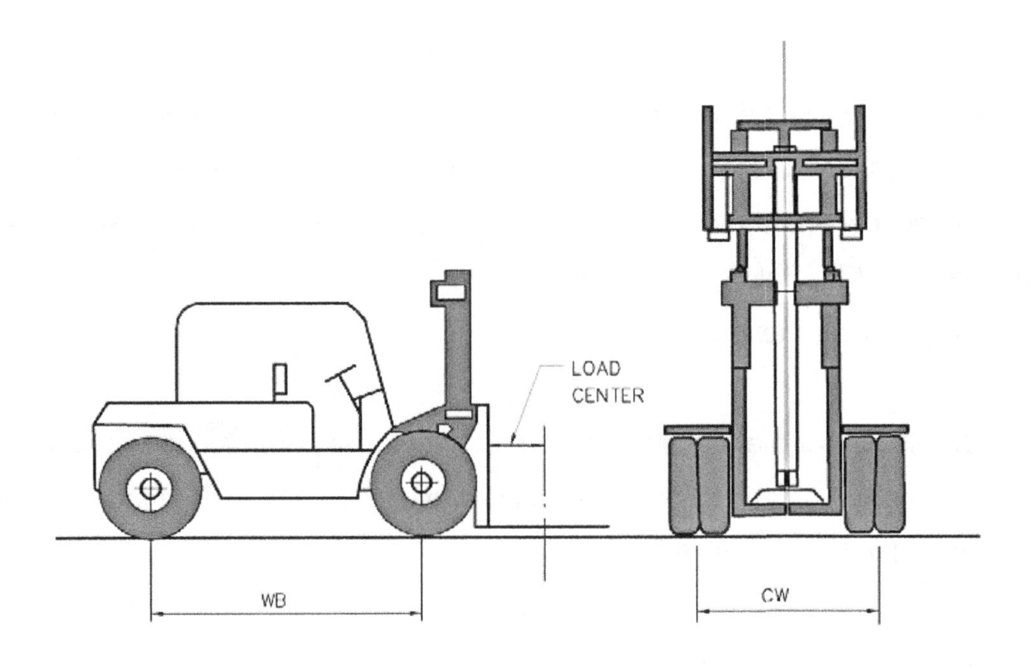

MAXIMUM LOAD (LBS)	LOAD CENTER (IN)	SERVICE WEIGHT (LBS)	TURNING RADIUS (FT.–IN)	WHEEL–BASE(WB) (FT.–IN)	WHEEL SPAC.(CW) (FT.–IN)	WHEEL LOADS (LOADED)	
						EACH REAR SINGLE TIRE (LBS)	EACH FRONT DUAL TIRE (LBS)
10,000	24	15,000	12–10	8–3	6–3	2,000	10,500
12,000	24	16,000	12–10	8–3	6–3	2,500	11,500
15,000	24	19,000	13–0	8–9	6–4	2,500	14,500
16,000	24	19,500	13–0	8–9	6–4	2,500	15,250
20,000	24	20,000	14–0	9–6	6–4	2,500	17,500
24,000	24	25,300	14–9	10–0	6–4	2,500	22,150
30,000	24	34,000	15–3	10–9	6–6	3,000	29,000
40,000	36	63,000	14–11	10–0	8–0	2,500	49,000

Figure 3-4
Wheel loads for forklifts

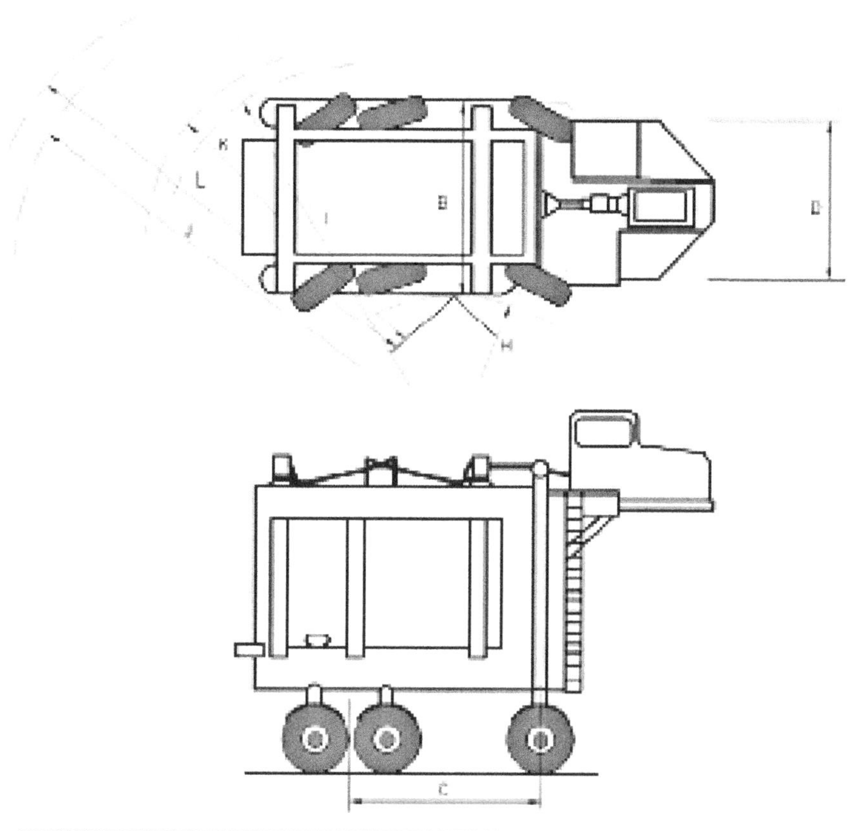

DEAD WEIGHT		67,000 LBS
SERVICE WEIGHT		89,000 LBS
WHEEL LOAD–EACH		26,000 LBS
OVERALL WIDTH	(B)	13'-4"
WHEELBASE	(C)	13'-4"
WHEEL CENTERS	(D)	11'-6"
INSIDE TURNING RADIUS	(H)	11'-3"
OUTSIDE TURNING (20ft. CONTAINER)	(I)	27'-10"
OUTSIDE TURNING (40ft. CONTAINER)	(J)	33'-4"
MINIMUM (20ft. CONTAINER)	(K)	19'-4"
MINIMUM (40ft. CONTAINER)	(L)	25'-4"

Figure 3-5

Wheel loads for straddle carriers

3.10.2 SIMULTANEOUS LOADS. Generally, apply uniform and concentrated live loads in a logical manner. Designated uniform live loadings and concentrated live loading from pneumatic-tired equipment should not be applied simultaneously in the same area. However, uniform live load should be assumed between crane tracks (for 80 percent of gage). When railroad tracks are present between crane tracks, both track loads should be applied simultaneously. However, the maximum loads from each track need not be assumed. Unique operations may warrant a more conservative approach, i.e, during a "trans-shipment operation (handling of fuel containers), there could be a portal crane straddling a railroad car loaded with a container, with another container sitting on the pier deck in the immediate vicinity.

3.10.3 SKIP LOADING. For determining the shear and bending moments in continuous members, apply the designated uniform loadings only on those spans that produce the maximum effect.

3.10.4 CRITICAL LOADINGS.

Concentrated loads from trucks, mobile cranes, forklifts, and straddle carriers, including mobile crane float loads, are generally critical for the design of short spans such as deck slabs and trench covers. Uniform loading, mobile crane float loading, rail-mounted crane loading, and railroad loading are generally critical for the design of beams, pile caps, and supporting piles.

4 HORIZONTAL LOADS

4.1 BERTHING LOADS. Procedures for calculation of berthing loads are found in the technical literature.

4.2 MOORING LOADS. Procedures for calculation of mooring loads are found in the technical literature.

4.3 WIND LOADS ON STRUCTURES. Use Section 1609 (or its current equivalent) of the International Building Code to provide minimum wind load on structures.

4.4 LOADS ON PILES. In addition to the axial loads, bending moments, and shears caused by lateral loads at deck level due to berthing, mooring and seismic forces, piles are also subjected to other types of lateral loads acting along the length of the pile.

4.4.1 CURRENT AND WAVES. These loads are applied at and near the water level and may be significant where large size piles are used in high-current waters. An estimate of current and wave forces can be made using methods in the technical literature.

4.4.2 SLOPING FILL LOADS. These loads are transmitted along the shaft of the piles by the lateral movement of the soil surrounding the piles beneath the structure, such as may occur along a sloping shoreline at marginal wharves. The maximum moments in the piles for this category of loadings are determined by structural analysis and the methods outlined in the technical literature after the conditions of pile support in the pile cap and the soil have been established and the effective length of pile has been determined.

- Piles of relieving platform types of solid wharves, shown on Figure 2-10, may be subjected to lateral earth loads if the stability of the slope beneath the platform is minimal and soil creep occurs. In such cases, stabilizing measures should be introduced, prior to installation of piles, to prevent movement of the soil along the slope. Among the stabilizing measures that may be used are surcharging (preloading), installation of sand drains or soil compaction piles, or replacement of unstable materials. If the piles supporting the structure are used to increase slope stability, or if time-dependent stabilizing measures are introduced after the piles are in place, calculate the resistance to soil movement provided by the piles and the piles checked for the bending moments induced by the calculated lateral earth loads, in addition to the increased loading caused by the downdrag of the settling soil.

- The pile resistance to soil movement may be obtained from a stability analysis by determining the additional resistance, provided by the piles, which will provide a factor of safety that corresponds to zero soil movement. The minimum factor of safety required for this tape of analysis varies and should be selected after evaluating the soil conditions, which exist at the site. The embedment length of piles needed for developing the required lateral resistance may be determined in accordance with the criteria given in UFC 3-220-01N.

4.4.3 PILE DRIVING LOADS. Piles are subjected to high compressive and tensile stresses during driving and should be proportioned to resist these in addition to the service loads. Where prolonged driving in alternately soft and hard layers of soil or driving through stiff "quaky" clays is anticipated, very high tensile stresses are set up and will require a higher level of prestress (1000 psi (6894.8 kPa) or more) in prestressed concrete piles. Give attention to controlling driving stresses by specifying frequent cushion replacement, and by requiring use of hammers capable of adjusting driving energy.

4.5 EARTHQUAKE LOADS.

4.5.1 DETERMINE SEISMIC USE GROUP:

- SUG III: Structures having significant importance and where the impact of being out of operation is critical (High Risk).
- SUG II: Structures having moderate importance where the impact of being out of operation for repairs is tolerable (Moderate Risk).
- SUG I: Structures not falling into SUG III, or II (Low Risk).

4.5.2 DETERMINE SEISMIC DESIGN CATEGORY: (Based on International Building Code, 2003, Table 1616.3)

4.5.3 DETERMINE SEISMIC ANALYSIS PROCEDURE:

	SDC A	SDC B	SDC C	SDC D	SDC E	SDC F
SUG I	Note A	Note A	Note A	Note A	Note B	Note B
SUG II	Note A	Note A	Note A	Note B	Note B	Note B
SUG III	Note A	Note A	Note B	Note B	Note B	Note B

Note A: Performance Based Seismic Design not required. Lateral loads from berthing and mooring of vessels will likely govern the design. Determine seismic forces utilizing the Equivalent Lateral Force procedure from SEI/ASCE 7-02, Minimum Design Loads for Buildings and Other Structures, paragraph 9.5.5 and utilizing the Design earthquake motion corresponding to Seismic Performance Level 2 as defined in Table 4-2 of Marine Oil Terminal Engineering and Maintenance Standards, June 2003, California State Lands Commission Marine Facilities Division.

Note B: Performance Based Seismic Design required according to the provisions of Marine Oil Terminal Engineering and Maintenance Standards, June 2003, California State Lands Commission Marine Facilities Division.

4.5.4 PERFORMANCE BASED SEISMIC DESIGN PROVISIONS.

a. Earthquake Loads: Follow the provisions of *Marine Oil Terminal Engineering and Maintenance Standards*, Section 3.4. The spectral acceleration values for earthquakes with a 2 and 10 percent probability of exceedance in a 50-year period are typically well defined. To obtain seismic spectral accelerations for other recurrence intervals use the following formula based upon the procedure presented in Section 1.6.1.3 of FEMA 356, *Prestandard and Commentary for the Seismic Rehabilitation of Buildings*.

Equation 3-1

$$S_i = S_{i(10/50)} (C_i),$$

where:

S_i = Spectral acceleration parameter("i" = "s" for short-period, or "i" = "1" for 1 second period)

$S_{i(10/50)}$ = Spectral acceleration parameter("i" = "s" for short-period, or "i" = "1" for 1 second period) at a 10% probability of exceedance in 50 years

C_i = Modifier provided in Tables 3-4 and 3-5 below ("i" = "s" for short-period, or "i" = "1" for 1 second period).

Table 3-4 Values of C_s, Modifier for Short-Period Spectral Acceleration

Design Earthquake Probability of Exceedance in 50 Years

Region	10 %	15 %	20 %	50 %	65 %	75 %
California	1.0	0.83	0.72	0.44	0.36	0.32
Pacific Northwest	1.0	0.79	0.67	0.36	0.29	0.25
Eastern US	1.0	0.72	0.56	0.23	0.17	0.14
Other locations	1.0	0.79	0.67	0.36	0.29	0.25

Notes:

1. The modifier C_s includes a statistical adjustment factor, "P_R", and a location adjustment factor, "n", as defined in section 1.6.1.3 in FEMA 356, *Prestandard and Commentary for the Seismic Rehabilitation of Buildings*, Nov. 2000. The value of "n" is taken as 0.54 for locations outside of the continental US.

2. This table is valid at locations where the Mapped MCE spectral acceleration for short-periods, S_S, is less than 1.5g.

Rehabilitation of Buildings, Nov. 2000. The value of "n" is taken as 0.54 for locations outside of the continental US.

2. This table is valid at locations where the Mapped MCE spectral acceleration for short-periods, S_S, is less than 1.5g.

Table 3-5 Values of C_1, Modifier for 1-Second Period Spectral Acceleration

Design Earthquake Probability of Exceedance in 50 Years

Region	10 %	15 %	20 %	50 %	65 %	75 %
California	1.0	0.83	0.72	0.44	0.36	0.32
Pacific Northwest	1.0	0.77	0.64	0.33	0.26	0.22
Eastern US	1.0	0.71	0.55	0.22	0.16	0.13
Other locations	1.0	0.77	0.64	0.33	0.26	0.22

Notes:

1. The modifier C_1 includes a statistical adjustment factor, "P_R", and a location adjustment factor, "n", as defined in section 1.6.1.3 in FEMA 356, *Prestandard and Commentary for the Seismic Rehabilitation of Buildings*, Nov. 2000. The value of "n" is taken as 0.59 for locations outside of the continental US.

2. This table is valid at locations where the Mapped MCE spectral acceleration for short-periods, S_S, is less than 1.5g.

b. Load Combinations: combine earthquake forces with other loadings according to paragraph 3-5 of this UFC.

c. Seismic Analysis and Structural Performance: Follow the provisions of *Marine Oil Terminal Engineering and Maintenance Standards*, June 2003, California State Lands Commission Marine Facilities Division, Section 4, as modified below.

Replace "MOT" with "Piers and Wharf Structures"

Replace Section 4.1.3 with the following: "Piers and Wharves shall be categorized into one of three risk classifications (high, moderate or low) according to SUG.

SUG III = High Risk Classification

SUG II = Moderate Risk Classification

SUG I = Low Risk Classification

Retitle section 4.2 "Existing and New Piers and Wharves".

Delete section 4.3 in its entirety.

Structural Analysis and Design of Components: Follow the provisions of *Marine Oil Terminal Engineering and Maintenance Standards*, June 2003, California State Lands Commission Marine Facilities Division, Section 7 as modified below.

Replace "MOT" with "Piers and Wharf Structures"

4.5.5 DYNAMIC FILL LOADS. In general, piles subjected to seismic forces behave as flexible members and their behavior is controlled primarily by the surrounding soil. Both vertical and batter piles move together with the surrounding soil during an earthquake. Provided that shear failure or liquefaction of the surrounding soil does not occur during ground shaking, the pile-supported structure will move a limited amount and remain stable after an earthquake. The magnitude of the horizontal movement depends on the earthquake magnitude and duration, design details of the platform, flexibility of the piles, and the subgrade modulus of the foundation soil. If the soil surrounding the piles is susceptible to liquefaction or if slope failure occurs, the piles will move excessively, resulting in serious damage to the piles and the structure. For these conditions, remove and replace unstable materials. When the piles penetrate a deep soft layer first and then a stiff layer of soil, the soils displace cyclically back and forth during an earthquake. During the cyclic ground shaking, the piles will move with the ground and return essentially to their original position if the soil does not fail during these cyclic displacements. Accordingly, if piles are to continue to safely support loads after an earthquake, it will be necessary for the piles to have the capability to withstand the induced curvature without failure.

4.5.6 EMBANKMENTS AND FILLS. For determining the stability of embankments and fills at solid wharves, when subjected to earthquake forces, refer to the technical literature.

4.5.7 FLOATING STRUCTURES. Usually, floating structures are not directly affected by seismic events. However, waves created by offshore seismic activity such as a seiche or tsunami will affect floating structures. Also, the mooring system employed (spud piles and chain) will be subjected to the ground motions and should be investigated.

4.6 EARTH AND WATER PRESSURES ON RETAINING STRUCTURES.

4.6.1 STATIC CASE. Static earth pressures, acting on retaining structures, are determined in accordance with the criteria detailed in the technical literature.

4.6.2 DYNAMIC CASE. Seismic forces may cause increased lateral earth pressures on earth retaining wharf structures accompanied by lateral movements of the structure. The degree of ground shaking that retaining structures will be able to withstand will depend, to a considerable extent, on the margin of safety provided for static loading conditions. In general, wharf retaining structures, designed conservatively for static loading conditions, may have a greater ability to withstand seismic forces than those designed, more economically, by less conservative procedures. Methods for determining lateral earth pressures due to seismic forces are discussed in the technical literature.

4.6.3 WATER PRESSURE. Consider pressures due to water level differentials, resulting from tidal fluctuations and/or groundwater accumulations, in the design of sheet pile bulkheads, cells, and curtain walls, and in stability investigations for embankments and fills. Consider additional loading due to hydrodynamic pressure for retaining structures.

4.7 THERMAL LOADS.

4.7.1 TEMPERATURE DIFFERENTIAL. The effect of thermal forces that build up in the structure due to fluctuations in temperature will vary from those measured at the time of construction. For piers and wharves, the large body of adjacent water has a substantial moderating effect on the structure. Consequently, the structure may not attain an overall temperature 10 deg. F (-12.2 deg C) to 20 deg. F (-6.7 deg C) higher or lower than the water temperature. The effect will be even less for ballasted deck construction. However, unballasted decks may see a large temperature differential through depth. Solid-type piers and wharves and floating structures are less likely to be affected by temperature variations.

4.7.2 PILE-SUPPORTED STRUCTURES. Typically, decks of pile-supported structures will be subjected to temperature differential. However, since the axial stiffness of the deck elements will be much higher than the flexural stiffness of piles, the deck will expand or contract without any restraint from piles (for narrow marginal wharves, the short inboard piles may offer some restraint, and hence need to be analyzed) and will subject the piles to bending moments and shear forces. Locate batter piles so as not to restrain thermal motion (usually in the middle portion of a long structure).

4.8 ICE FORCES. In addition to the weight of accumulated ice on the structure, consider the forces exerted by floating ice. The principal modes of action of floating ice are shown in Figure 3-6 and discussed below.

4.8.1 DYNAMIC IMPACT. Follow the criteria in the AASHTO standard to the extent feasible. For lightly loaded structures and for open pile platforms, these criteria may result in structures of unreasonable proportions. In such cases, consider reducing the AASHTO criteria in accordance with the Canadian code. See Dynamic Ice Forces on Piers and Piles. The values of effective pressure are:

- AASHTO .... 400 psi (2757.9 kPa);
- Canadian Code .... 100 to 400 psi (689.5 to 2757.9 kPa) (highway bridges); and
- Canadian Code .... 200 to 250 psi (1379.0 to 1723.7 kPa) (wharf piles).

NO.	DESCRIPTION	TYPICAL ENVIRONMENT	ILLUSTRATION
1	IMPACT OF MOVING SHEETS AND FLOES.	RIVERS AT BREAK-UP. COASTAL WATERS WITH APPRECIABLE CURRENTS	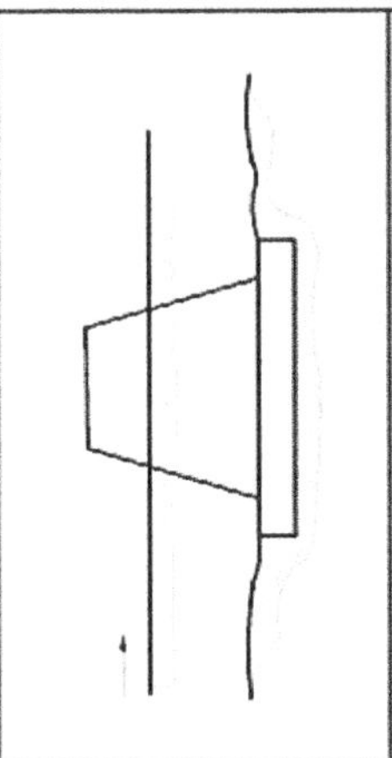
2	STATIC PRESSURE FROM EXPANDING OR CONTRACTING SHEETS.	LAKES, SHELTERED COASTAL WATERS. TEMPERATURE CHANGES AND JACKING BY REFREEZING OF CRACKS.	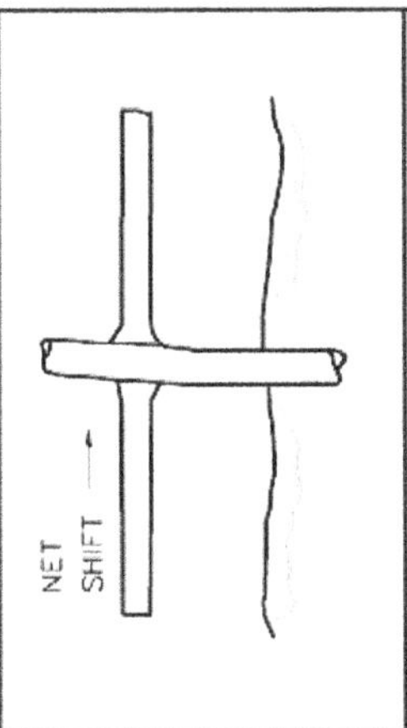 NET SHIFT →
3	SLOW PRESSURE FROM ICE PACK OR JAM.	EXPOSED COASTAL WATERS, RIVERS.	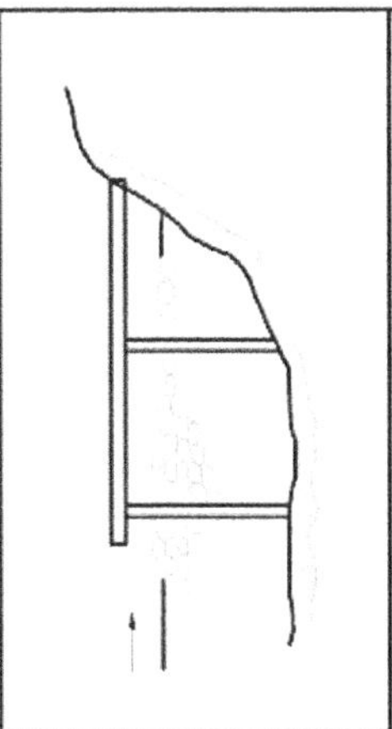
4	VERTICAL MOVEMENT	TIDAL LOCATIONS WITH HEAVY ICE BUILD-UP.	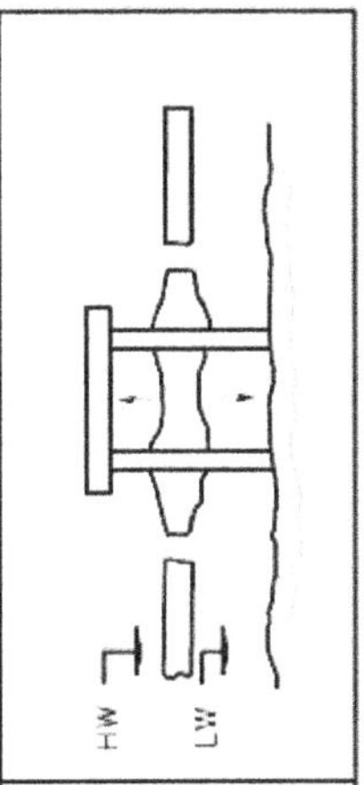 HW LW

Figure 3-6

Principal modes of ice action

4.8.2 STATIC PRESSURE. Freshwater ice will exert less pressure on a structure than seawater ice of the same thickness. For freshwater ice, assume pressures of 15 to 30 psi (103.4 to 106.8 kPa). For sea ice, pressures of 40 psi (275.8 kPa) to as much as 150 psi (1034.2 kPa) may be assumed. These are maximum values and relate to crushing of the ice. If the ice sheet can ride up on the nearby shore, the pressure exerted will be less than if the ice sheet is confined within vertical boundaries.

4.8.3 SLOW PRESSURE. Broken ice floes will exert less pressure than a solid ice sheet. In general, the pressures developed in this mode of action will be less than those to be experienced under the static pressure mode of action. Reliable values of pressure are not presently available.

4.8.4 VERTICAL MOVEMENT. Assume that the structure will lift or depress a circular sheet of ice. Calculate the radius of the affected ice sheet on the basis of the flexural strength of ice as 80 to 200 psi (1379.0 kPa.) Check the shear on the basis of the strength (and adhesion) as 80 to 150 psi (551.6 to 1034.2 kPa.) Consider the formation of bustle (added thickness) of ice around the structure.

4.9 SHRINKAGE. Open pier and wharf decks, which are usually constructed from concrete components, are subject to forces resulting from shrinkage of concrete from the curing process. Shrinkage loads are similar to temperature loads in the sense that both are internal loads. For long continuous open piers and wharves and their approaches, shrinkage load is significant and should be considered. However, for pile-supported pier and wharf structures, the effect is not as critical as it may seem at first because, over the long time period in which the shrinkage takes place, the soil surrounding the piles will slowly "give" and relieve the forces on the piles caused by the shrinking deck. The PCI Design Handbook is recommended for design.

4.10 CREEP. This is also a material-specific internal load similar to shrinkage and temperature and is critical only to prestressed concrete construction. The creep effect is also referred to as rib shortening and should be evaluated using the PCI Design Handbook.

5. LOAD COMBINATIONS. Proportion piers and wharves to safely resist the load combinations represented by Tables 3-6 and 3-7. Analyze each component of the structure and the foundation elements for all the applicable combinations. Tables 3-6 and 3-7 list load factors (f) to be used for each combination and the percentage of unit stress applicable for service load combinations.

Equation 3-2

$$S_l \text{ or } U_l = f_D (D) + f_L (L_c + I \text{ or } L_u) + f_{Be} (Be) + f_B (B)$$
$$+ f_C (C) + f_E (E) + f_{Eq} (Eq) + f_W (W) + f_{Ws} (Ws)$$
$$+ f_{RST} (R + S + T) + f_{Ice} (Ice))$$

where:

S_l = Service load combination

U_l = Ultimate load combination

f_x = Load factor listed in Tables 3-6 and 3-7

3-5.2 LOAD SYMBOLS. The following load symbols are applicable for Equation 3-1:

D = Dead load

L_u = Live load (uniform)

L_c = Live load (concentrated)

I = Impact load(for Lc only)

B = Buoyancy load

B_e = Berthing load

C = Current load on structure

C_s = Current load on ship

E = Earth pressure load

EQ = Earthquake load

W = Wind load on structure

W_s = Wind load on ship

R = Creep/rib shortening

S = Shrinkage

T = Temperature load

Ice = Ice pressure

5.3 LOAD FACTOR DESIGN. Concrete structures for piers and wharves may be proportioned using the Load Resistance Factor Design (ultimate strength) method; however, they should be checked for serviceability and construction loads. Load combinations are presented in Table 3-6.

5.4 SERVICE LOAD DESIGN. Timber structures for piers and wharves should be proportioned using the service load combinations and allowable stresses. Concrete and steel structures may also be designed using the above approach. The service load approach should also be used for designing all foundations and for checking foundation stability. Load combinations are presented in Table 3-7.

VACANT	1(a)	2(b)	3(c)	4(d)	5(e)	6(f)	7(g)	8(h)
D	1.4	1.2	1.2	1.2	1.2	1.2	0.9	0.9
L	0	\1\ 1.6[1] /1/	1	0	1	1	0	0
B	1.4	1.2	1.2	1.2	1.2	1.2	0.9	0.9
Be	0	0	0	0	0	0	0	0
C	1.4	1.2	1.2	1.2	1.2	1.2	0.9	0.9
Cs	0	0	0	0	0	0	0	0
E	0	1.6	0	0	0	0	1.6	1.6
EQ	0	0	0	0	0	1	0	1
W	0	0	0	0.8	1.6	0	1.6	0
Ws	0	0	0	0	0	0	0	0
RST	0	1.2	0	0	0	0	0	0
Ice	0	0.2	0	0	1	0	1	0

BERTHING	1(a)	2(b)	3(c)	4(d)	5(e)	6(f)	7(g)	8(h)
D		1.2	1.2		1.2	1.2		
L		\1\ 1.6[1] /1/	1		1	1		
B		1.2	1.2		1.2	1.2		
Be		1.6	1		1	1		
C		1.2	1.2		1.2	1.2		
Cs		0	0		0	0		
E		1.6	0		0	0		
EQ		0	0		0	1		
W		0	0		1.6	0		
Ws		0	0		0	0		
RST		1.2	0		0	0		
Ice		0.2	0		1	0		

MOORING	1(a)	2(b)	3(c)	4(d)	5(e)	6(f)	7(g)	8(h)
D	1.4	1.2	1.2	1.2	1.2	1.2	0.9	0.9
L	0	\1\ 1.6[1] /1/	1	0	1	1	0	0
B	1.4	1.2	1.2	1.2	1.2	1.2	0.9	0.9
Be	0	0	0	0	0	0	0	0
C	1.4	1.2	1.2	1.2	1.2	1.2	0.9	0.9
Cs	1.4	1.2	1.2	1.2	1.2	1.2	0.9	0.9
E	0	1.6	0	0	0	0	1.6	1.6
EQ	0	0	0	0	0	1	0	1
W	0	0	0	0.8	1.6	0	1.6	0
Ws	0	0	0	0.8	1.6	0	1.6	0
RST	0	1.2	0	0	0	0	0	0
Ice	0	0.2	0	0	1	0	1	0

\1\ [1] 1.3 for maximum outrigger float load from a truck crane /1/

(a) *SEI/ASCE 7-02 Minimum Design Loads for Buildings and Other Structures*, 2.3.2 Equation 1

(b) *SEI/ASCE 7-02 Minimum Design Loads for Buildings and Other Structures*, 2.3.2 Equation 2

(c) *SEI/ASCE 7-02 Minimum Design Loads for Buildings and Other Structures*, 2.3.2 Equation 3a

(d) *SEI/ASCE 7-02 Minimum Design Loads for Buildings and Other Structures*, 2.3.2 Equation 3b

(e) *SEI/ASCE 7-02 Minimum Design Loads for Buildings and Other Structures*, 2.3.2 Equation 4

(f) *SEI/ASCE 7-02 Minimum Design Loads for Buildings and Other Structures*, 2.3.2 Equation 5

(g) *SEI/ASCE 7-02 Minimum Design Loads for Buildings and Other Structures*, 2.3.2 Equation 6

(h) *SEI/ASCE 7-02 Minimum Design Loads for Buildings and Other Structures*, 2.3.2 Equation 7

Table 3-6

Load Combinations – LRFD

VACANT	1(a)	2(b)	3(c)	4(d)	5(e)	6(f)	7(g)	8(h)	9(i)	10(j)
D	1	1	1	1	1	1	1	1	0.6	0.6
L	0	1	0	0.75	0	0	0.75	0.75	0	0
B	1	1	1	1	1	1	1	1	0.6	0.6
Be	0	0	0	0	0	0	0	0	0	0
C	1	1	1	1	1	1	1	1	0.6	0.6
Cs	0	0	0	0	0	0	0	0	0	0
E	0	1	1	1	1	1	1	1	1	1
EQ	0	0	0	0	0	0.7	0	0.525	0	0.7
W	0	0	0	0	1	0	0.75	0	1	0
Ws	0	0	0	0	0	0	0	0	0	0
RST	0	1	0	0.75	0	0	0	0	0	0
Ice	0	0.7	0.7	0	0	0	0	0	0.7	0

BERTHING	1(a)	2(b)	3(c)	4(d)	5(e)	6(f)	7(g)	8(h)	9(i)	10(j)
D		1		1			1	1		
L		1		0.75			0.75	0.75		
B		1		1			1	1		
Be		1		0.75			0.75	0.75		
C		1		1			1	1		
Cs		0		0			0	0		
E		1		1			1	1		
EQ		0		0			0	0.525		
W		0		0			0.75	0		
Ws		0		0			0	0		
RST		1		0.75			0	0		
Ice		0.7		0			0	0		

MOORING	1(a)	2(b)	3(c)	4(d)	5(e)	6(f)	7(g)	8(h)	9(i)	10(j)
D	1	1	1	1	1	1	1	1	0.6	0.6
L	0	1	0	0.75	0	0	0.75	0.75	0	0
B	1	1	1	1	1	1	1	1	0.6	0.6
Be	0	0	0	0	0	0	0	0	0	0
C	1	1	1	1	1	1	1	1	0.6	0.6
Cs	1	1	1	1	1	1	1	1	0.6	0.6
E	0	1	1	1	1	1	1	1	1	1
EQ	0	0	0	0	0	0.7	0	0.525	0	0.7
W	0	0	0	0	1	0	0.75	0	1	0
Ws	0	0	0	0	1	0	0.75	0	1	0
RST	0	1	0	0.75	0	0	0	0	0	0
Ice	0	0.7	0.7	0	0	0	0	0	0.7	0

(a) *SEI/ASCE 7-02 Minimum Design Loads for Buildings and Other Structures*, 2.4.1 Equation 1
(b) *SEI/ASCE 7-02 Minimum Design Loads for Buildings and Other Structures*, 2.4.1 Equation 2
(c) *SEI/ASCE 7-02 Minimum Design Loads for Buildings and Other Structures*, 2.4.1 Equation 3
(d) *SEI/ASCE 7-02 Minimum Design Loads for Buildings and Other Structures*, 2.4.1 Equation 4
(e) *SEI/ASCE 7-02 Minimum Design Loads for Buildings and Other Structures*, 2.4.1 Equation 5a
(f) *SEI/ASCE 7-02 Minimum Design Loads for Buildings and Other Structures*, 2.4.1 Equation 5b
(g) *SEI/ASCE 7-02 Minimum Design Loads for Buildings and Other Structures*, 2.4.1 Equation 6a
(h) *SEI/ASCE 7-02 Minimum Design Loads for Buildings and Other Structures*, 2.4.1 Equation 6b
(i) *SEI/ASCE 7-02 Minimum Design Loads for Buildings and Other Structures*, 2.4.1 Equation 7
(j) *SEI/ASCE 7-02 Minimum Design Loads for Buildings and Other Structures*, 2.4.1 Equation 8

Table 3-7

Load Combinations - Allowable Stress Design

CHAPTER 3
STRUCTURAL DESIGN

1. CONSTRUCTION MATERIALS.

1.1 TIMBER. For the major functional types of piers and wharves such as berthing, repair, fitting-out/refit, and supply facilities subject to high concentrated wheel loads, timber construction should not be considered. Timber may be more effective and optimal for fender systems, dolphins, walkways, utility trays, and deck-supported small buildings. For light-duty piers and wharves, such as fueling, temporary, and Magnetic Treatment and Electromagnetic Roll piers, timber framing for deck and piling may be considered. Consult with the local activity and Facilities Engineering Command (FEC) for local requirements and restrictions on the use of treated timber. Obtaining permits for treated timbers, especially creosoted fender pile, is becoming very difficult in various areas of the country.

1.1.1 PRESERVATIVE TREATMENT. All timber members, with the exception of some fender piling, exposed to the marine environment and immersed in salt water or fresh water should be pressure treated with oilborne (creosote, pentachlorophenol) or waterborne (salts) chemical preservatives to protect against deleterious effects of decay, insects, and marine borers. In warmer waters, where severe marine borer activity can be anticipated, employ dual treatment using both creosote and salt. Consult the staff entomologist at the cognizant FEC for specific information on marine organisms present and the treatment required. If possible, make pressure treatment after all holes and cuts are made. When holes and cuts are made in the field, treat timber members with preservative to prevent decay from starting in the holes or cuts. Consult a staff entomologist for the proper preservative. Field treatments are difficult in the tidal zone and are typically not very effective against marine borers. Therefore, whenever possible, design and detailing should avoid the necessity for making cuts or holes on underwater timber members. For example, avoid bracing or connections below mean high water. All connection hardware should be suitable for the saltwater exposure. For above water construction, waterborne salt treatment is preferable to creosote treatment due to the staining effect that creosote treatment produces.

1.1.2 TIMBER SPECIES. Douglas fir and southern pine are the most popular species for waterfront construction. Southern pine piles are limited to 65 ft (19.8 m) in length, whereas Douglas fir piles and poles can be used in up to 100-ft (30.5 m) lengths. Large beams and timber sizes needed for chocks and wales are generally available only in Douglas fir and southern pine. Treat chocks and wales with waterborne salts and not oilborne preservatives such as creosote. Evaluate the cost and availability of timber piles and other members for the project.

1.2 STEEL. When protected against corrosion by the use of coal tar epoxy or other marine coatings and cathodic protection systems, steel construction may be considered for all types of marine structures. Active cathodic systems are difficult to design, construct, and maintain properly, therefore, passive systems are preferred. Additional steel thickness may be provided as a sacrificial corrosion allowance. Steel is particularly adaptable for use: in template and jack-up barge construction at advance base facilities; as piles for structures located in deep water where high lateral forces must be resisted; as fender piles and fender panels; as piles for structures located in areas of high seismic activity; and where difficult driving is anticipated. When the utilization of other construction materials is considered feasible, the use of steel construction may be restricted due to cost and maintenance considerations.

1.3 CONCRETE. For piers and wharves, concrete is generally the best material for construction. Properly designed and constructed facilities are highly durable in the marine environment. New advances in concrete technology have improved concrete durability. Concrete enhanced with fly ash, silica fume and corrosion inhibitors has demonstrated superior performance and should be used whenever possible. Concrete is immune to marine borer and insect attack and is incombustible. Concrete is also ideal for deck construction in open-type piers and wharves and, when properly designed, is more economical for floating structures. Proprietary stainless steel reinforcement bars, wires, and strands have been developed for use in concrete construction where nonmagnetic properties are desired as in Magnetic Treatment and Electromagnetic Roll piers.

1.3.1 PRECAST CONCRETE PILES. Precast concrete piles should preferably be prestressed to resist the tensile forces frequently encountered during driving. Corrosion of reinforcement in prestressed concrete piles can be controlled by proper mix design and, in extreme cases, by epoxy coating the reinforcement. However, exercise sufficient control during driving of concrete piles to minimize cracking. Where difficult driving into very compact sands, gravels, or rock is anticipated, the tip of the piles may be equipped with a WF-shape or H-pile "stinger" to achieve needed penetration. Very large hollow cylindrical piles (48-in. diameter and larger) have also been successfully employed for waterfront construction on the East coast. However, longitudinal cracking was encountered with these same types of piles during construction of a pier at Bremerton, Washington.

1.4 COMPOSITES. Composites made of concrete and steel, concrete and fiberglass, plastic and fiberglass, and plastic and steel have been successfully employed in piers and wharves. Composites offer many advantages over conventional materials but often have limitations that need to be considered. Some advantages may include improved corrosion resistance, lightweight, and ease of construction. Some of the disadvantages may include low strength, UV light deterioration, long-term durability and high cost.

1.4.1 CONCRETE AND STEEL. Concrete-filled pipe piles, steel H-piles with a concrete casing, and steel beams with concrete decks are the more common composite types. The concrete casing or jacket for the steel H-piles may be required only in the splash or tidal zone. However, due to instances of severe deterioration of the pile just below the jacket, the jacket must be extended to ELW or to below mudline. Concrete may also be added to steel pipe piles for deadweight purposes to resist uplift forces or to increase the stiffness of the pile.

1.4.2 CONCRETE AND FIBERGLASS. Concrete-filled fiberglass piles have been used in facilities where high axial capacities are not required. The lightweight fiberglass piles are easily installed and do not require high capacity handling equipment.

1.4.3 PLASTIC AND FIBERGLASS. Fiberglass reinforced plastic piles and beams have been successfully used in pier and wharf construction primarily as fender piles, wales and chocks.

1.4.4 PLASTIC AND STEEL. Recycled plastic piles with steel cage reinforcement have also been used in pier and wharf construction.

1.5 ALUMINUM. For deck-supported structures and for support of piping and conduits, aluminum members are useful. However, do not use unprotected aluminum underwater or in the splash zone. To prevent corrosion, aluminum should be electrically isolated from adjacent materials by nonconductive gaskets, washers, or bolt sleeves. Aluminum construction may be used in the superstructure of Magnetic Treatment and Electromagnetic Roll piers, due to the nonmagnetic characteristics of the material.

1.6 PLASTICS. Fiberglass-reinforced plastics (FRP), ultra-high molecular weight (UHMW) plastics, and high-density polyethylene (HDPE) are being increasingly used in waterfront construction. FRP grating and shapes are highly durable in the marine environment when shielded from ultraviolet rays. UHMW plastics are useful in fender systems design as rub strips where a high abrasion resistance and low coefficient of friction are required. UHMW plastics are available in various grades. The use of corrosion-resistant fiber reinforced plastic (FRP) components including reinforcing bars, prestressing tendons, structural shapes, and unidirectional or woven fabrics, are being developed and have been successfully used in the repair of piers and wharves. Consider using these materials when the situation warrants, but special attention must be given to the design of connections. Carefully evaluate the use of FRP components as structural elements for new construction.

2. ALLOWABLE STRESSES

2.1 GENERAL. Allowable stresses for materials used in pier and wharf construction generally conform to industry standard codes for the type of material and the purposed application unless modified herein.

2.2 TIMBER. Design timber structural elements in accordance with the technical literature. Allowable stresses are generally not affected by preservative treatment. However, modulus of rupture and modulus of elasticity are considerably reduced by preservative treatment. When preservative treatment for fire retardation is used, the allowable stresses should be reduced by 10 percent.

2.3 STEEL. Design steel structural elements in accordance with applicable codes, standards and best practices.

2.4 CONCRETE. Design concrete structural elements in accordance with applicable codes, standards, and best practices. For prestressed concrete members, "zero" tension design is preferred. All reinforced concrete deck members should meet the crack control requirements for severe exposure.

2.5 OTHER MATERIALS. Fiberglass-reinforced plastics (FRP), ultra-high molecular weight (UHMW) plastics, and other new materials should be governed by the accepted industry standards for structural design and detailing.

3. DECK STRUCTURE DESIGN

3.1 DECK FRAMING. Concrete is generally considered the best material for deck framing and should be used for most pier and wharf decks. Although timber, steel, steel/concrete composite, and timber/concrete composite decks have been used in the past, they are neither cost-effective nor suitable for the high concentrated load capacities currently demanded of decks. From durability, maintenance, and life-cycle-cost viewpoints, a concrete deck is superior and is highly recommended. The deck framing should be slabs supported on pile caps, using an all cast-in-place, all precast, or composite construction, as shown in Figure 4-1. For the concentrated loads which typically control the deck design, a solid slab with its high punching shear resistance is recommended. Framing systems using thin slabs, as in cast-in-place slab/beam/girder systems, should not be used because of the tendency to spall along beam/girder corners and edges. Occasionally, where high concentrated loads are not specified, voided slabs may be used. Map cracking in the cast-in-place topping at the precast panel joints is sometimes seen. To control the cracking, transverse post-tensioning is sometime utilized. For distribution of horizontal loads, pier and wharf decks should be continuous, with as few expansion joints as possible. Where expansion joints are needed, the deck on each side of the joint should be supported on a separate pile cap or girder.

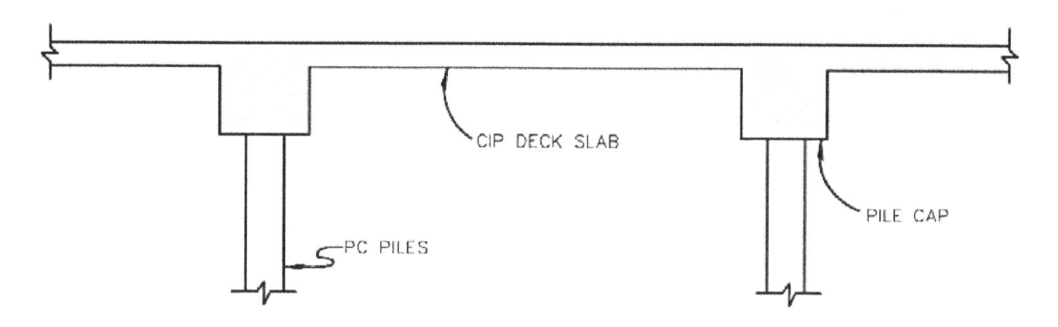

(A) ALL CAST-IN-PLACE DECK

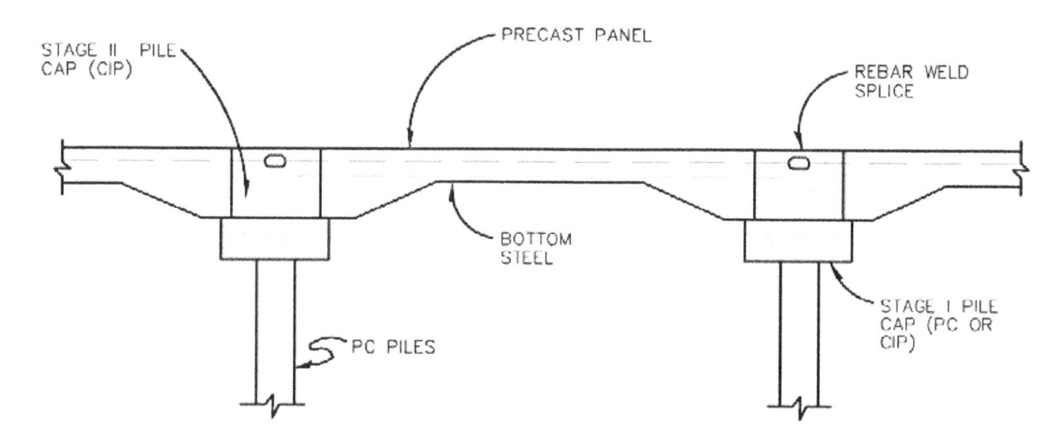

(B) ALL PRECAST DECK

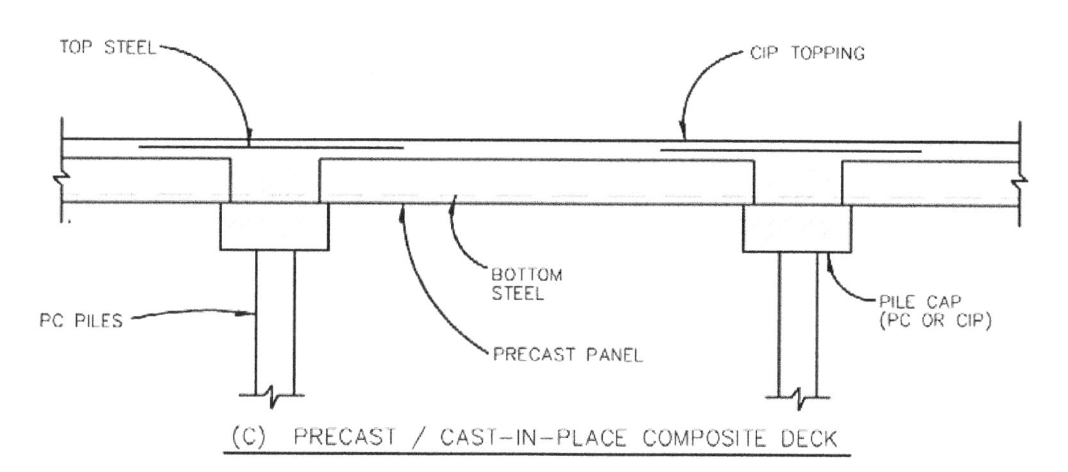

Figure 4-1

Concrete Deck Construction

3.2 PILE CAPS. It is often cost-effective to orient pile caps (and hence pile bents) transverse to the length of the structure. This orientation provides improved lateral stiffness for berthing and mooring forces. When this orientation is used, longitudinal pile caps are not needed unless crane trackage support or longitudinal seismic resistance is to be provided. For marginal wharves where lateral loads from mooring and berthing loads are transferred to the land, a longitudinal orientation of the pile cap may be considered if feasible for construction. Moments and shears on pile caps from live loads should take into account the elastic shortening of the piles and the effect of soil deformation at and near pile tips. For computation of forces from high concentrated loads, the cap behaves as a beam on elastic foundation, and distributes the concentrated load to a number of piles adjacent to the load. While hand calculations are acceptable, a stiffness analysis using a computer is recommended.

4. SUBSTRUCTURE DESIGN

4.1 PILE BENT FRAMING. A pile-supported framing system is the most popular form for substructure design for open piers and wharves. Several framing concepts for open piers and wharves and marginal wharves are illustrated in Figure 4-2. Many variations and combinations of the illustrated concepts are possible.

4.1.1 ALL PLUMB PILE SYSTEM. The lateral loads are resisted by "frame action," whereby the piles and the cap form a moment frame and resist the lateral load primarily by the flexural stiffness of the piles. However, for narrow structures, lateral deflection may be high for even small lateral loads. Also, sidesway is not prevented, which increases the effective length of the pile as a column. If piles vary in unsupported length, the shorter piles will attract a large portion of the lateral load. Because the piles are more efficient for axial loads and less so for bending moments, this framing usually is restricted to shallow waters and light lateral loads. However, for wide structures with a large number of piles, the total stiffness of the system may justify a reduced effective length. A more in-depth stability analysis is needed to validate a reduced effective length. Large diameter steel pipe and precast/prestressed concrete cylinder piles can provide improved lateral stiffness and are attractive for use in areas of high seismic activity.

4.1.2 PLUMB/BATTER PILE SYSTEMS. In this type of framing, all the vertical loads are primarily handled by the plumb piles, and lateral loads are resisted primarily by the batter piles. The behavior of the system is one of "truss action." This system is more cost-effective as the lateral loads are resisted primarily by the axial stiffness of the batter piles. However, very high forces are transmitted to the caps, which will have to be designed and detailed to resist these forces. In areas of high seismic activity, the increased stiffness of the system reduces the period and leads to higher earthquake loads.

4.1.3 ALL BATTER PILE SYSTEM. This system is a compromise between the two above, and is cost-effective in some circumstances. With this system, the batter slope may be near vertical. Natural periods can be as high as several seconds, making the approach attractive for seismic areas.

4.1.4 BATTER PILE SYSTEM WITH SEISMIC ISOLATION. This system incorporates calibrated isolators or seismic fuses between the wharf deck and batter piles. The system allows for high displacements of the wharf deck once a threshold lateral load causes the isolator slip. Consider the magnitude of lateral berthing and mooring forces such that they do not exceed the threshold lateral force of the isolator. In this case a separate fendering structure may be required.

5. MOORING HARDWARE

5.1 GENERAL. Ships are usually moored to bollards, bitts, and cleats. Occasionally, ships may be tied to a quick-release hook. The position of a ship on a berth is usually controlled by utility hookup and brow location requirements. The crew in charge of tying up the ship will usually tie the lines to whatever mooring hardware is convenient to give the required horizontal angle. This often results in lines tied to a lower capacity cleat while a high-capacity bollard may only be a few feet away. Hence, consider using only one type of high-capacity mooring hardware throughout the facility. When possible, size this mooring hardware for the maximum size vessel that could possibly use the facility. Space hardware to maximize the berth flexibility for use by ships other than the specific vessel the berth was designed to accommodate. In addition, mooring hardware requirements will depend on the mooring service type assigned to the berth. For mooring service types III and IV, consider providing additional heavy weather mooring hardware. Balance the desire to provide the higher capacity hardware with the additional cost of the higher strength hardware and supporting structure. The geometry of the hardware should preclude mooring lines from slipping off, as the mooring angle is often very steep.

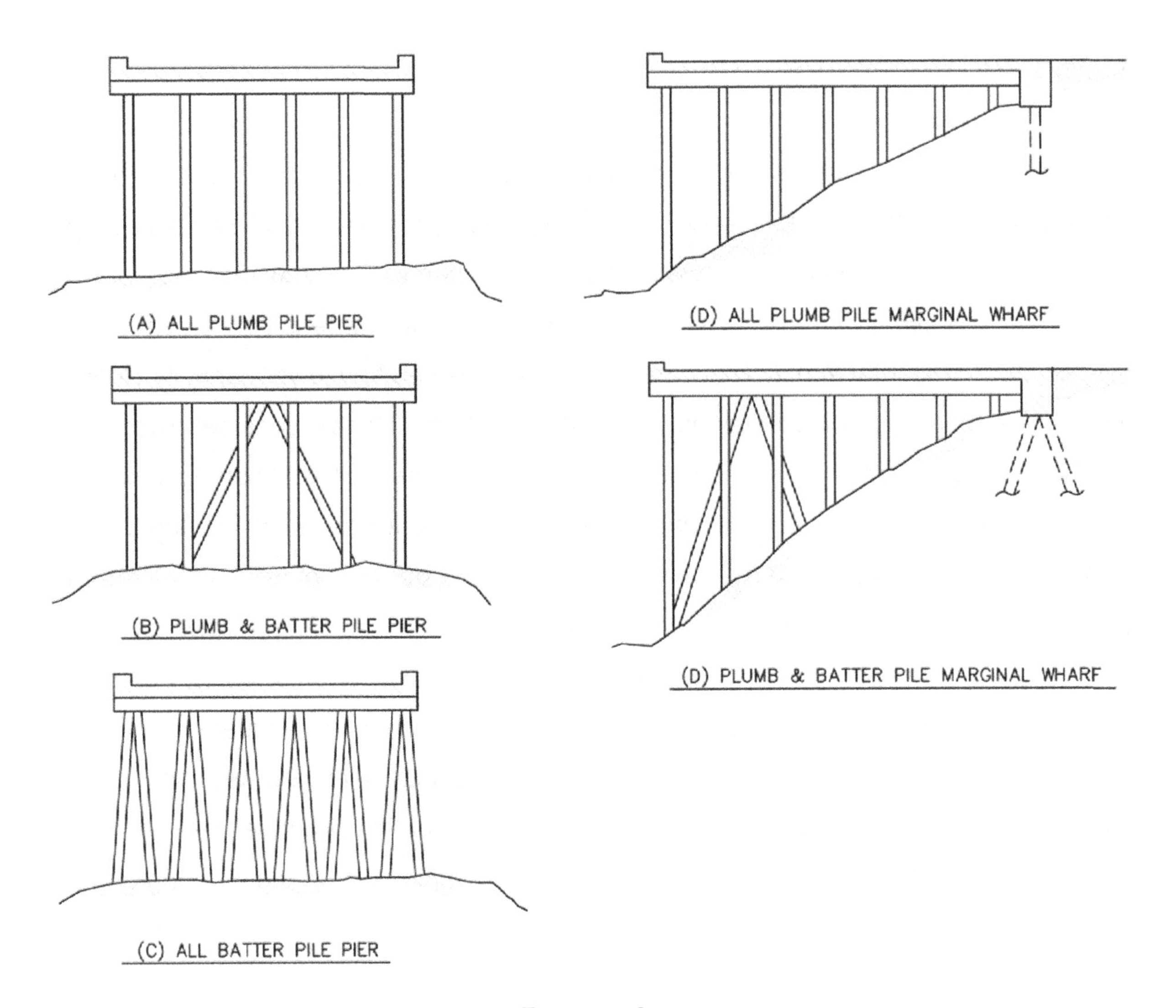

Figure 4-2

Substructure Framing Concepts

5.2 HARDWARE TYPES. The following are the most commonly used types of mooring hardware:

5.2.1 BOLLARDS. A bollard is a short single-column cast-steel fitting that extends up from a baseplate that is secured to a strong point of a shore structure or berthing facility. Bollards are used in snubbing or checking the motion of a ship being moored, by tightening and loosening mooring lines that are fastened to them. Bollards are also used for securing a ship that has been placed in its final moored position. Do not use bollards without ears in facilities where a high vertical angle of the mooring line is anticipated, to prevent lines from slipping off.

5.2.2 BITTS. Bitts are short, double-column, cast-steel fittings fastened to the deck of berthing facilities. They are used to snub and secure a vessel. The double-columns allow for convenient and rapid tying and releasing of mooring lines, as well as for guiding a line through to other hardware.

5.2.3 CLEATS. Currently, available cleats are low-capacity, cast steel deck fittings having two projecting arms that are intended to be used for securing mooring lines of small craft. They are provided at many facilities. Given a choice, line-handling crews will use cleats in preference to bollards or bitts, even for large ships, as the possibility of line slippage is remote. However, cleats can easily be overloaded when they are used in lieu of major fittings such as bollards. Because of the low holding capacity of cleats, they should not be used in combination with higher capacity deck fittings.

5.2.4 CHOCKS. Chocks are either stationary or roller-equipped cast deck fittings that are used to train the direction of a mooring line. Chocks are available either open at the top, permanently closed, or closed by a hinged closing piece.

5.2.5 CAPSTANS. Ships outfitted with winch-mounted wire rope mooring lines require greater pulling power than can be provided by one or two deck hands to draw out the ship's lines. The assignment is handled by capstans mounted along the face of the wharf. The capstans are small electric winches of 5 to 10 hp with a drum rotating about a vertical axis. The capstan is used by a deck hand that receives a messenger line at the end of which is fastened the sling of the wire rope hawser. The capstan, receiving several winds of the messenger line, provides the pulling power needed to draw out the wire rope hawser. The messenger line is then returned to the ship. Capstans are also used as primary guidance (breasting and in-haul) to berth ships in drydocks and slip-type berths (Trident facilities). For these uses, the capstans are of larger capacity and are typically two-speed.

5.2.6 QUICK-RELEASE HOOKS. The quick-release hook, generally mounted on a swivel base, is a deck fitting used to receive mooring lines. When a ship is required to make a hasty departure from its berth, a tug on the hook's release mechanism unfastens the mooring line. The mechanism can also be tripped from the ship when a tag line is provided. Thus, a ship can make a sudden departure without the assistance of shore personnel. Quick-release mooring hooks with integral power capstans are necessary for securing the steel mooring lines on petroleum tankers at fuel piers, while bollards are needed for the supplementary lines other than steel.

5.3 STRENGTH. The required strength of mooring hardware and its fastenings is determined by the breaking strength of the strongest mooring line or lines that may be fastened to it. Mooring hardware can and does receive more than one line and as many as three are not unusual. The sizes of mooring lines are limited to those that can be conveniently handled by deck hands. Thus, wire rope lines generally do not exceed \1\ 1-3/4 in (44.5 mm) /1/ diameter.

5.4 PLACEMENT. If a berthing facility were always to receive the same class of ship, each of which had identical arrangements for putting out mooring lines, a specific pattern for mooring hardware spacing, based on the ship's fittings, would be satisfactory. However, most naval berthing facilities require a high degree of flexibility in order to be able to receive several types and sizes of ships. Therefore, a universal pattern for mooring hardware spacing is preferred. Mooring hardware spaced at 60 ft on centers along the berthing face of a structure will, in most instances, provide the number of fittings required to secure the ships during the periods of time that wind velocities and conditions of sea do not exceed the design criteria established for mooring service types I and II. Mooring service types III and IV will likely require additional high capacity storm bollards which are normally set back at least 100 feet from the face of the berth to provide a shallow line angle.

- A berthing facility that will accommodate ships having large wind presentments, such as aircraft carriers, should be outfitted with 12 – 100 ton bollards at 100-ft (30.5 m) centers and 4 – 200 ton storm bollards at each end. Locate the storm bollards, which would be used to secure breasting lines, inshore from the face of the wharf, thus reducing vertical angles and permitting the use of longer (safer) mooring lines.

- At submarine berths where mooring lines go down to the submarine, locate the mooring hardware as close as possible to the waterside edge of the bullrail to minimize chafing of the lines. Where this is not feasible, cast a continuous smooth member, such as a bent plate, in the concrete bullrail, as shown in Figure 4-3.

6 MOORING DOLPHINS/PLATFORMS

6.1 DESIGN. The primary load for a mooring dolphin comes from the tension in the mooring line. It is typically constructed as an open pile supported structure. Where filled (solid) construction is permitted, a single sheet pile cell may be used. When a platform is provided for the dolphin, it should be large enough to allow a 3-ft-wide (0.9 m) walking space all around the mooring hardware.

6.2 CONSTRUCTION. Timber mooring dolphins can be constructed from 7, 19, or 37 wood piles with a king pile in the center and other piles arranged in a circular pattern around the king pile. A 19-pile dolphin is illustrated in Figure 4-4(A). Limit the use of timber dolphins to facilities not requiring high capacity mooring points. Timber dolphins may be required for magnetic sensitive facilities. In areas with ready access to timber piles, timber dolphins may be popular for small craft, tugs, patrol craft and barges. For higher loads, a concrete dolphin is preferred, which is illustrated in Figure 4-4(B). Because both timber and concrete dolphins can be expected to move significantly (1 to 6 in (25 to 150 mm)) from the lateral load, design and detail the access walkway to allow for this movement. For situations where significant pile pullout capacities are needed, consider steel spin fin piles.

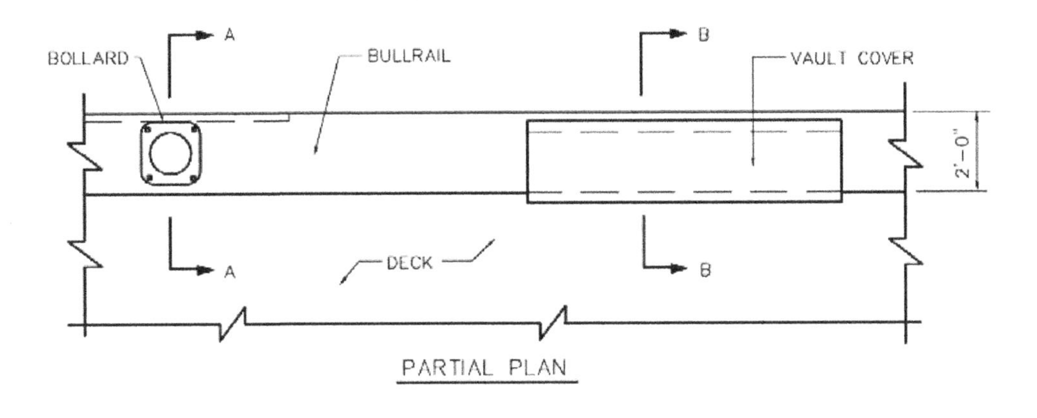

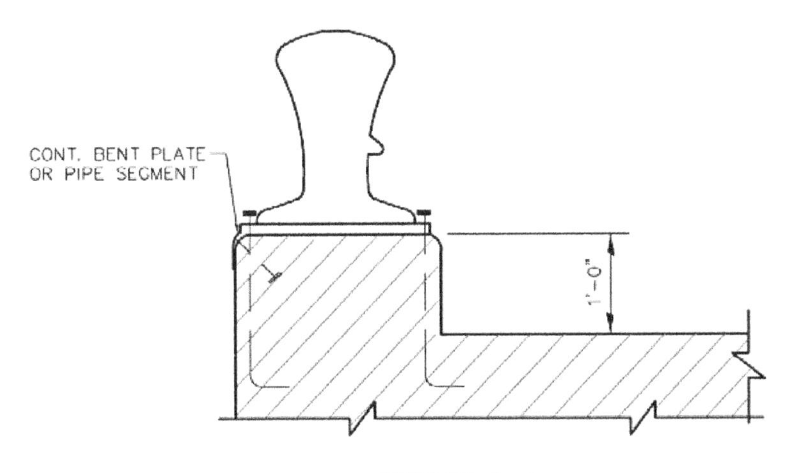

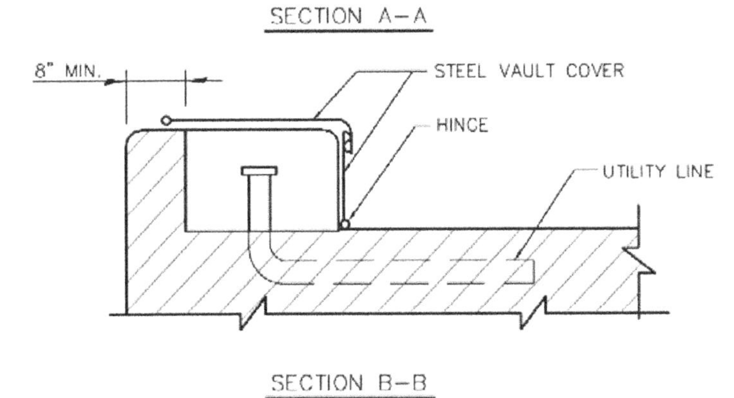

FIGURE 4-3

BULLRAIL DETAILS

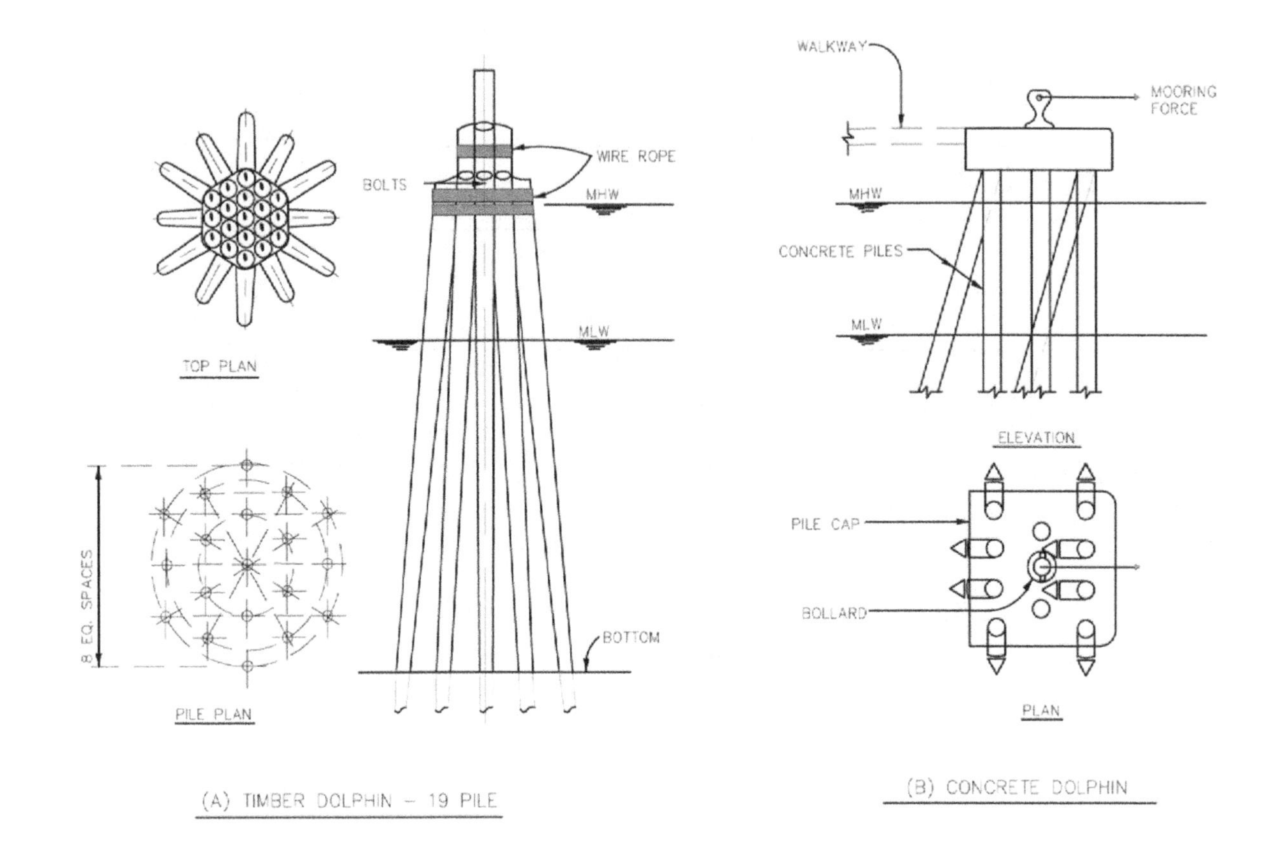

Figure 4-4
Mooring Dolphins

7. MISCELLANEOUS CONSIDERATIONS

7.1 EXPANSION JOINTS. Because expansion joints require frequent maintenance for proper functioning, piers and wharves should use as few joints as possible. The size and number will depend on the temperature range and structural system employed. Provide expansion joints at the junction of an approach with the main structure and such other places where there is a major structural discontinuity. Provide additional expansion joints where necessary to limit buildup of thermal loads. Continue the joint through railroad tracks and crane rail tracks. Recommended details are shown in Figures 4-5 and 4-6. Likewise, detail utilities crossing expansion joints to accommodate the expected longitudinal and lateral movements.

7.2 DRAINAGE. Pier and wharf decks should be sloped in transverse and longitudinal direction to deck drains or scuppers to provide for drainage of storm water. Where permitted, the storm water can be drained off to the water below; however, if fuel, oils, and chemicals are to be handled on the facility, the storm water should be collected and piped off for treatment. A complete review of local jurisdictional requirements for storm water management and treatment required for each facility may vary significantly between locations. It is customary to use subsurface drains in ballasted decks to handle any small amounts of water that may seep through the paving. This water is normally not collected.

7.3 BULLRAIL. On all waterside edges of piers and wharves, provide a curb or bullrail 10 to 12 in (254 to 305 mm) high by 12 to 24 in (3.5 to 610 mm) wide. Some mooring hardware may be accommodated within the 24-in (610 mm) width, thus permitting a clear inside face for easy snow removal and line handling. As shown in Figure 4-3, it is also generally possible to house utility vaults within the width of the bullrail. The bullrail should be sufficiently reinforced and anchored to the deck structure. When a continuous bullrail is available, it may be reinforced to serve as chord member for a structural diaphragm.

7.4 UTILITY TRENCH. Because the utility services are mostly needed along the pier or wharf edge, the main utility trenches on single deck piers should be kept close to the bullrail. The trench may be underhung or kept above, as shown in Figure 4-7. The trench covers should be removable and made of concrete, steel, or composite construction. Although the trench covers need not be watertight, use a good seal at joints to prevent accidental seepage of spilled liquids. Frequently spaced drains should be placed along the trench to prevent flooding. Provide adequate width and depth to allow access for maintenance of utilities.

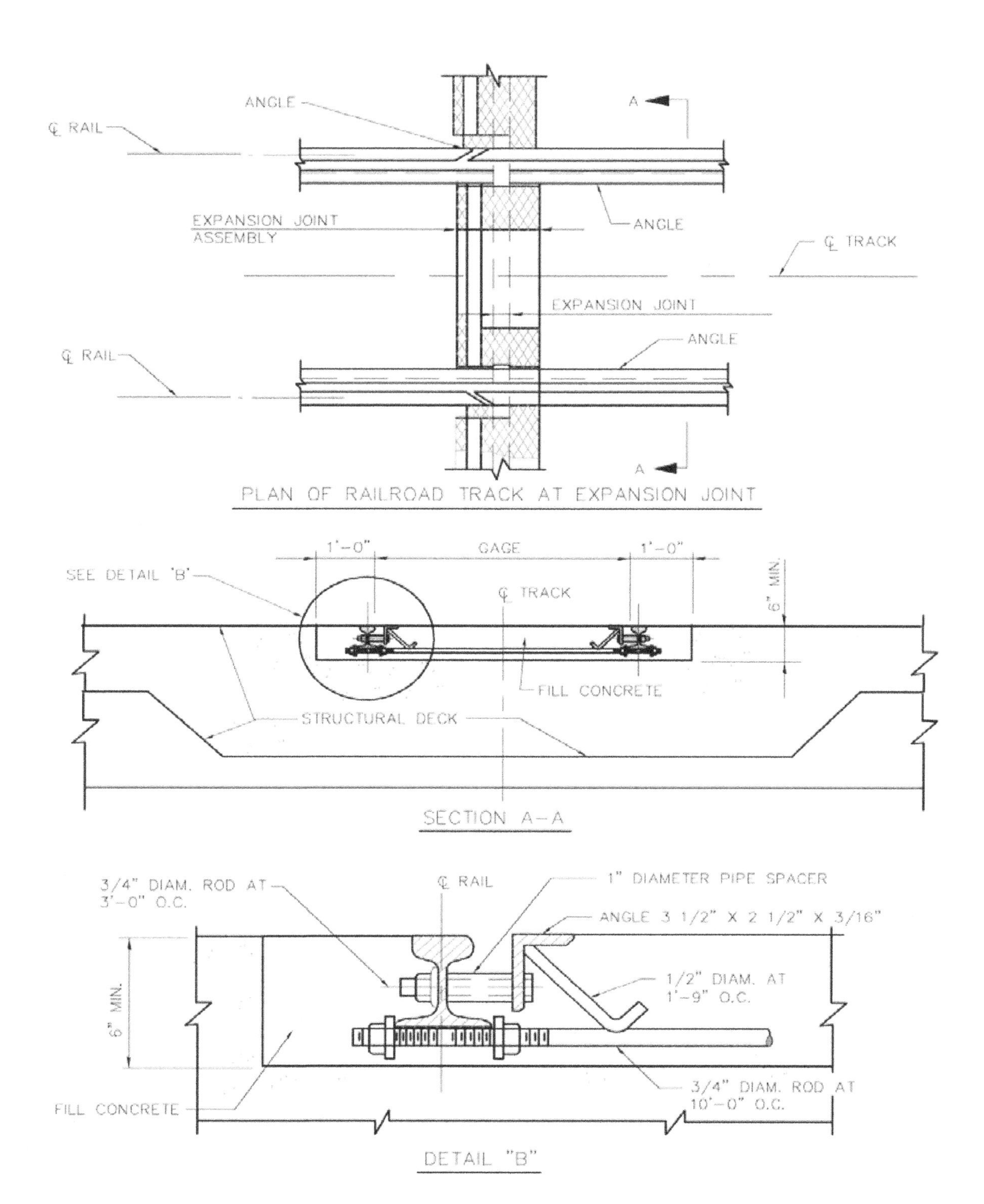

Figure 4-5

Railroad Track Support

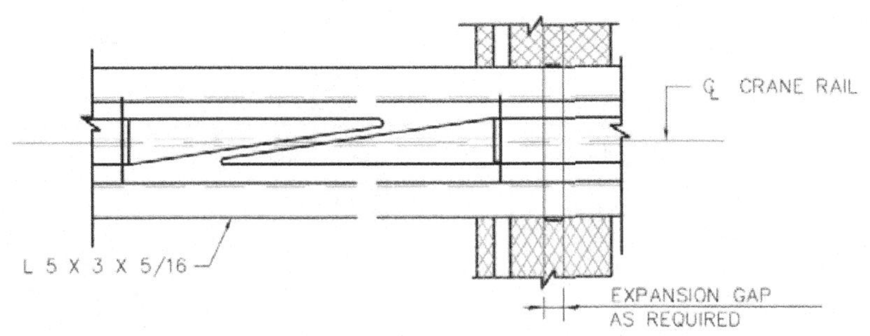

(A) PLAN OF CRANE RAIL AT EXPANSION JOINT

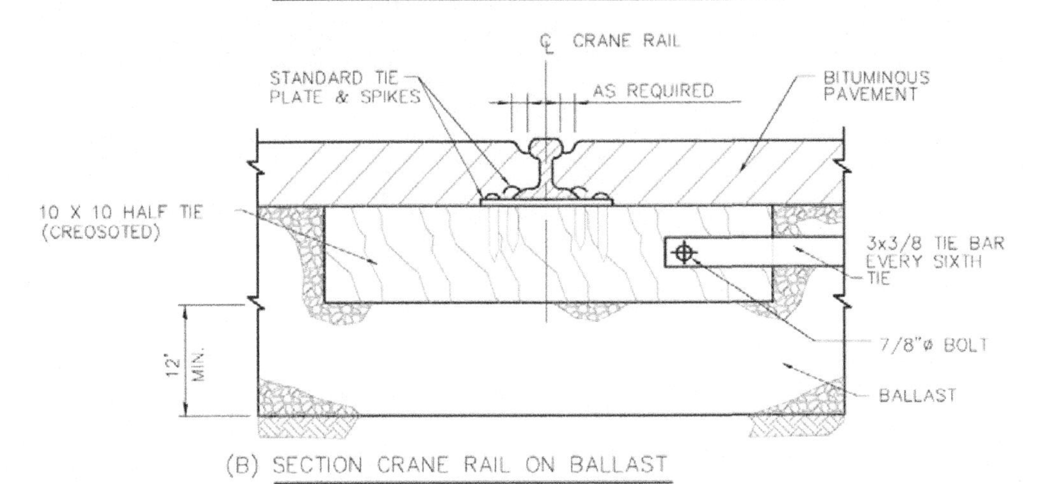

(B) SECTION CRANE RAIL ON BALLAST

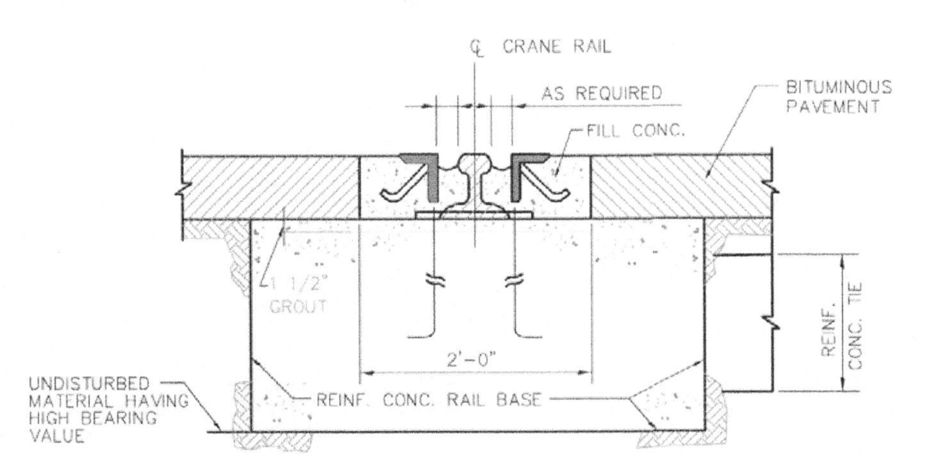

(C) SECTION CRANE RAIL ON CONCRETE BASE

Figure 4-6

Crane Rail Support

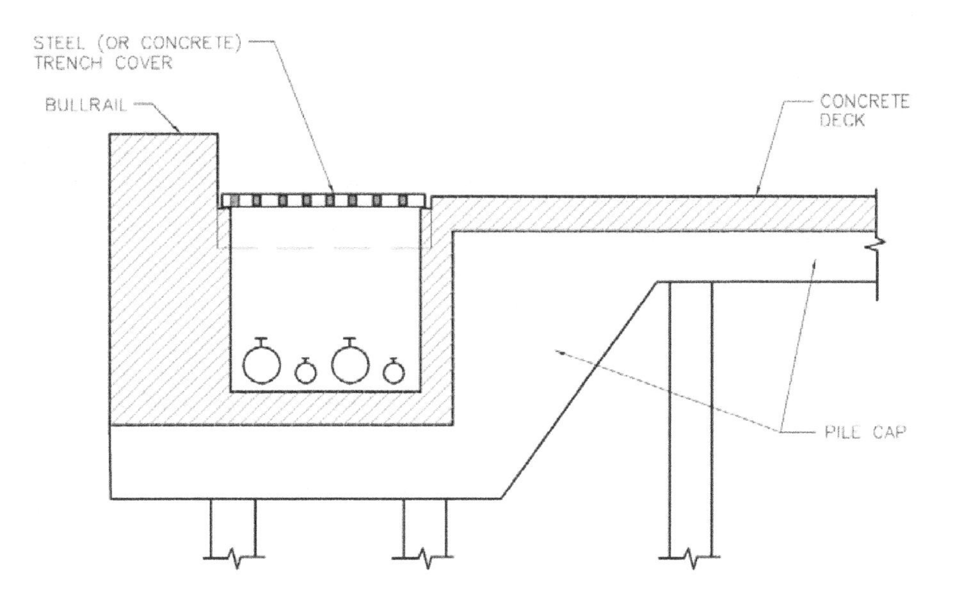

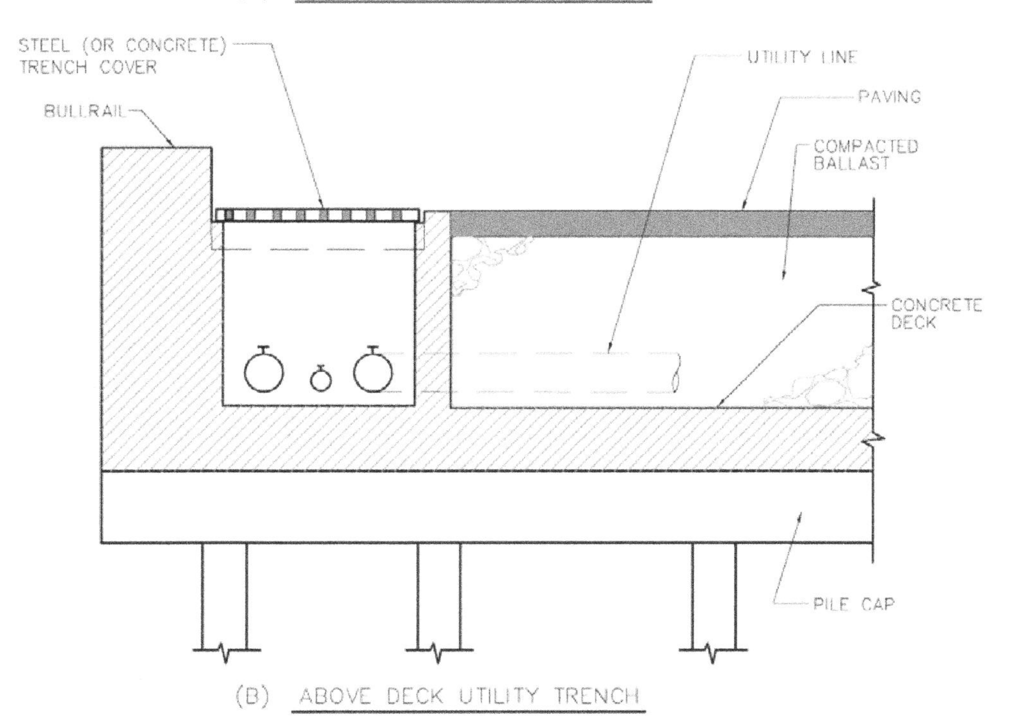

Figure 4-7

Utility Trench Concept

7.5 UTILITY HOODS. Protect connection points or "utility risers" projecting above deck or bullrail from snagging of mooring lines by pipe rails or concrete hoods, as shown in Figure 4-8. Alternatively, when the bullrail is wide enough, employ hinged utility covers like the one shown in Figure 4-3. For convenience of the users, the risers should not face away from the ship, and should be adequately sized for ease of operation.

7.6 WEARING COURSE. Generally, a wearing course is not needed for concrete deck structure. However, for precast concrete decks, an asphalt concrete or cement concrete wearing surface may be utilized to provide a slope for drainage. For ballasted decks and for solid (filled) type construction, a wearing course is necessary. For these applications, asphaltic concrete is preferred over cement concrete as the former is easier to repair and maintain and better tolerates substrate movement.

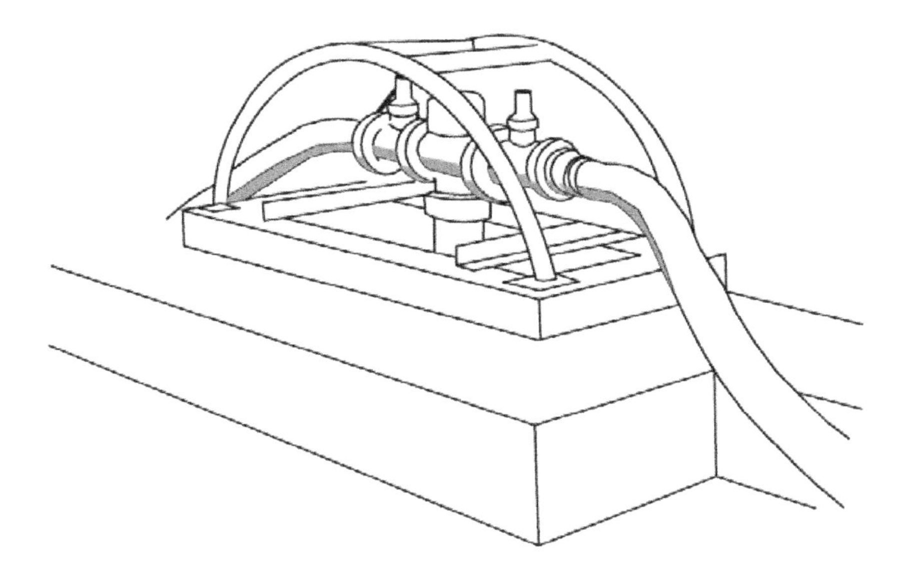

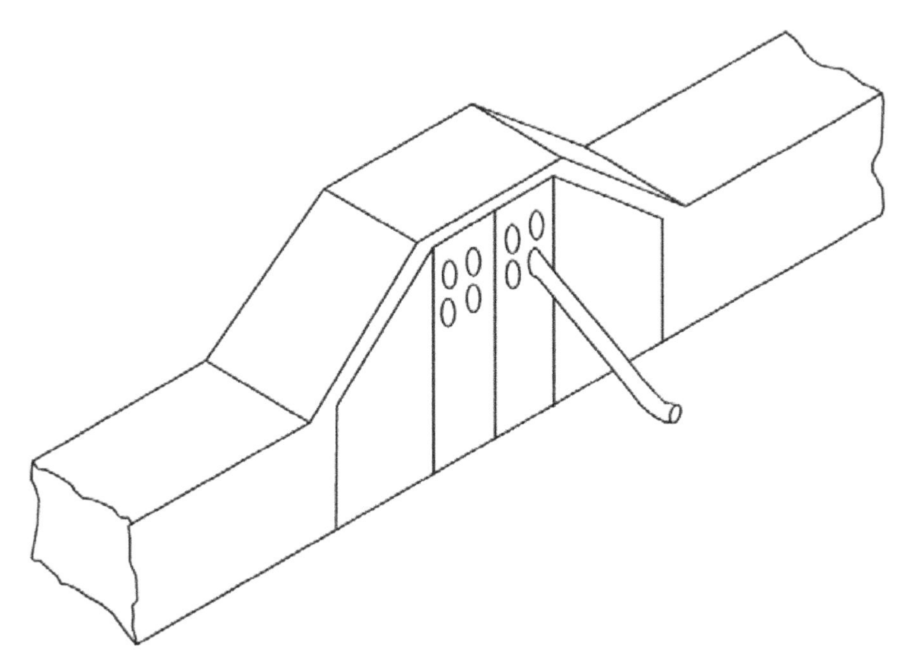

Figure 4-8

Utility Hoods

CHAPTER 4
FENDER SYSTEMS

1. GENERAL CONSIDERATIONS

1.1 DESCRIPTION. The fender system is the interface between the ship and the shore facility. During the berthing of a ship, the fender system is meant to act as a buffer in absorbing or dissipating the impact energy of the ship without causing permanent damage to the ship or the shore facility. Where ships are to be berthed against relatively inflexible solid piers and wharves, protection of the ship is a critical function. When ships are to be berthed against pile-supported piers, wharves, and dolphins (which are relatively flexible), protection of the structure may be the more serious concern. Once the ship is successfully berthed and moored to the shore facility, the fender system continues to provide the interface between ship and shore and transmits the environmental loads (wind, waves, and current) on the ship to the structure. For low-profile ship berthing, the fender system also provides a physical barrier to prevent the vessel from going underneath the pier.

1.2 BERTHING PRACTICE. The selection and design of a fender system is highly dependent on the berthing practice employed at the particular facility. Typically, two or more tugboats assist large ships into the berth. In some locations, smaller ships may be allowed to come in on their own power. When assisted by tugs, the ship would arrive off the berth and parallel to it. The ship then stops dead in the water and the tugs push and pull the ship transversely toward the berth in an attempt to make contact with as much of the fender system as possible. When unassisted by tugs, the smaller ship will be eased into its berth at some slight angle, referred to as the angle of approach. In both cases, the initial contact is limited to a relatively small portion of the fender system. Assumptions will have to be made regarding the approach angle and contact length.

1.3 CAMELS AND SEPARATORS. Camels are located between ships and piers or wharves. Separators are located between nested ships. This practice is the most significant difference between commercial ship berthing and naval ship berthing. Berthing against camels concentrates the impact energy to a small length of the fender system as well as applies the energy at some distance below the deck. This aspect must be recognized in all fender system design for ships. A fender system designed for commercial ships will, in general, not be satisfactory for military applications. The practice of using camels has resulted in a general trend for a minimum of hull protrusions near the waterline. Fender systems with higher fender contact area are more susceptible to damage from longitudinal movement of the vessel due to snags.

1.4 SYSTEMS APPROACH. The impact energy of the berthing ship is absorbed in a complex system of interconnected elements. For the system shown in Figure 5-1, the load passes from the ship's hull to the camel, which is backed by a series of fender piles. The fender piles, in turn, are supported by rubber fender units at the deck level. In this case, the ship's energy is absorbed by the ship's hull, rubbing strips, separator, fender piles, and rubber fenders at deck level. The system can be modeled as shown in Figure 5-1. The energy absorbed and the force developed, by each element can only be solved by an iterative process.

1.5 FUNCTIONAL REQUIREMENTS.

1.5.1 ENERGY ABSORPTION. Design all fender systems for absorption of the ship's berthing energy in all the structural types of piers and wharves.

1.5.2 NORMAL BERTHING. The fender system should be able to absorb the energy from normal berthing operations within the working stress or acceptable deformation range as defined in this section. Some manufacturers indicate a load deflection curve tolerance of plus or minus 10%. If this is determined to be the case, the design reaction on the structure should be increased by 10% and the energy absorption for design should be decreased by 10%. Variations in the speed of testing of fenders may affect the resulting load-deflection curves. Where the test loads are applied rapidly, i.e., at a speed comparable to the actual ship berthing, the load-deflection will indicate higher reaction and energy than if the test load is applied slowly. Therefore, take care when comparing test results from different manufacturers, and make appropriate adjustments in the factors of safety used in design. Differences of in the order of 30% can be expected.

1.5.3 ACCIDENTAL BERTHING. Because the fender system is less expensive than the ship or the berthing structure, some damage to it may be permissible and acceptable. So, in the event of an accidental situation, it is the fender system that should be "sacrificed." Loss of the berth has a much more serious consequence than loss of part or all of the fender system in terms of the cost and time required to restore the facility. The cost and time to repair a damaged ship is of much greater concern than the berth and the fender system. The accidental condition may be caused by increased approach angle or approach velocity or a unique situation that cannot be anticipated. In the absence of any other accident scenario, increase the berthing energy as calculated in this chapter by at least 50 percent and the fender system should be capable of providing this "reserve" capacity at or near failure of the system materials.

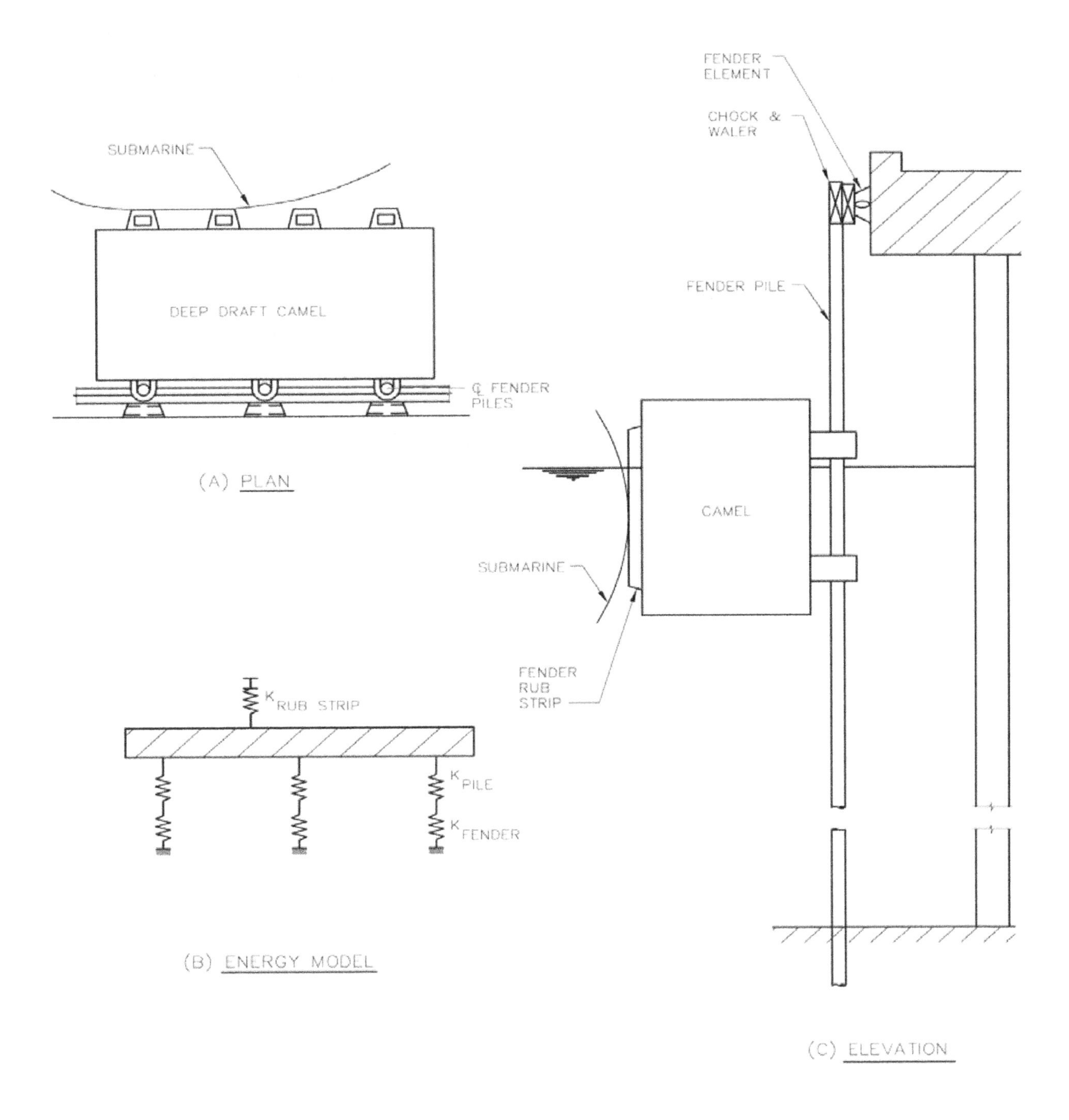

Figure 5-1

Energy model of a fender system

1.5.4 MOORED CONDITION. All fender systems selected should be capable of safely transferring the environmental loads on the ship to the mooring structure.

1.5.5 HULL DAMAGE. Design all fender systems to prevent permanent deformation of the ship's hull. It is much more expensive to repair a ship's hull than rehabilitate a damaged fender system. The composition of a typical hull is steel plating welded to longitudinal (horizontal) stiffeners at 2 to 4 ft (0.6 to 1.2 m) on center. The stiffeners span from 5 to 20 ft (1.5 to 7.6 m) depending on the vessel. Generally, the stiffeners are of sufficient strength to preclude failure from fender loading. However, the hull plating may yield when subjected to a uniformly distributed overload on the panel. Fender systems with rigid face elements or in combination with rigid camels tend to concentrate the reaction forces on the ships frames versus the hull plating due to the relative stiffness of the frames.

2. BERTHING ENERGY DETERMINATION.

2.1 METHODS. The following methods can be used in the determination of berthing energy of the ship.

2.1.1 KINETIC METHOD. The kinetic energy method has been the widely accepted method for piers and wharves of many facilities. When the displacement tonnage of the ship is known, the energy equation can be written as:

Equation 5-3

$$E_{ship} = \frac{1}{2}Wv^2/g$$

where

E_{ship} = Berthing energy of ship (ft-lbs)

W = Weight of the ship in pounds (displacement tonnage x 2,240)

g = Acceleration due to gravity (32.2 ft/second2)

v = Berthing velocity normal to the berth (ft/second)

However, there are several factors that modify the actual energy to be absorbed by the fender system. The expression can be written as

Equation 5-4

$$E_{fender} = C_b \times C_m \times E_{ship}$$

where:

E_{fender} = Energy to be absorbed by the fender system

C_b = Berthing coefficient = $C_e \times C_g \times C_d \times C_c$. Sometimes Eccentricity (C_e), geometric (C_g), deformation (C_d), and configuration (C_c) coefficients are combined into a single value called berthing coefficient.

C_m = Effective mass or virtual mass coefficient

Each of these coefficients is discussed separately below.

2.1.1.1 ECCENTRICITY COEFFICIENT (C_e). During the berthing maneuver, when the ship is not exactly parallel to the berthing line, not all the kinetic energy of the ship will be transmitted to the fenders. Due to the reaction from the fender, the ship will start to rotate around the contact point, thus dissipating part of its energy. Treating the ship as a rigid rod of negligible width in the analysis of the energy of impact on the fenders leads to the simple formula:

Equation 5-5

$$C_e = k^2/(a^2 + k^2)$$

where

k = Radius of longitudinal gyration of the ship, ft.

a = Distance between the ship's center of gravity and the point of contact on the ship's side, projected onto the ship's longitudinal axis, ft.

Values of Ce typically are between 0.4 and 0.7. The values for Ce may be computed from Figure 5-2.

2.1.1.2 GEOMETRIC COEFFICIENT (C_g). The geometric coefficient, Cg, depends upon the geometric configuration of the ship at the point of impact. It varies from 0.85 for an increasing convex curvature to 1.25 for concave curvature. Generally, 0.95 is ended for the impact point at or beyond the quarter points of the ship, and 1.0 for broadside berthing in which contact is made along the straight side.

2.1.1.3 DEFORMATION COEFFICIENT (C_d). This accounts for the energy reduction effects due to local deformation of the ship's hull and deflection of the whole ship along its longitudinal axis. The energy absorbed by the ship depends on the relative stiffness of the ship and the obstruction. The deformation coefficient varies from 0.9 for a non-resilient fender to nearly 1.0 for a flexible fender. For larger ships on energy-absorbing fender systems, little or no deformation of the ship takes place; therefore, a coefficient of 1.0 is recommended.

2.1.1.4 CONFIGURATION COEFFICIENT (C_c). This factor has been introduced to take into account the difference between an open pier or wharf and a solid pier or wharf. In the first case, the movements of the water surrounding the berthing ship are not (or hardly) affected by the berth. In the second case, the water between the berthing ship and the structure is squeezed, which introduces a cushion effect that represents an extra force on the ship away from the berth and reduces the energy to be absorbed by the fender system. Therefore, a reduction factor has to take care of this effect. Experience has indicated that for a solid quaywall about one quarter of the energy of the berthing ship is absorbed by the water cushion; therefore, the following values for C_c appear to be justified:

- For open berth and corners of solid piers, $C_c = 1.0$.
- For solid piers with parallel approach, $C_c = 0.8$.
- For berths with different conditions, C_c might be chosen somewhere between these values.

2.1.1.5 EFFECTIVE MASS OR VIRTUAL MASS COEFFICIENT (C_m). When a ship approaches a dock, the berthing impact is induced not only by the mass of the moving ship, but also by the water mass moving along with the ship. The latter is generally called the "hydrodynamic" or "added" mass. In determining the kinetic energy of a berthing ship, the effective or virtual mass (a sum of ship mass and hydrodynamic mass) should be used. The hydrodynamic mass does not necessarily vary with the mass of the ship, but is closely related to the projected area of the ship at right angles to the direction of motion. Other factors, such as the form of ship, water depth, berthing velocity, and acceleration and deceleration of the ship, will have some effect on the hydrodynamic mass.

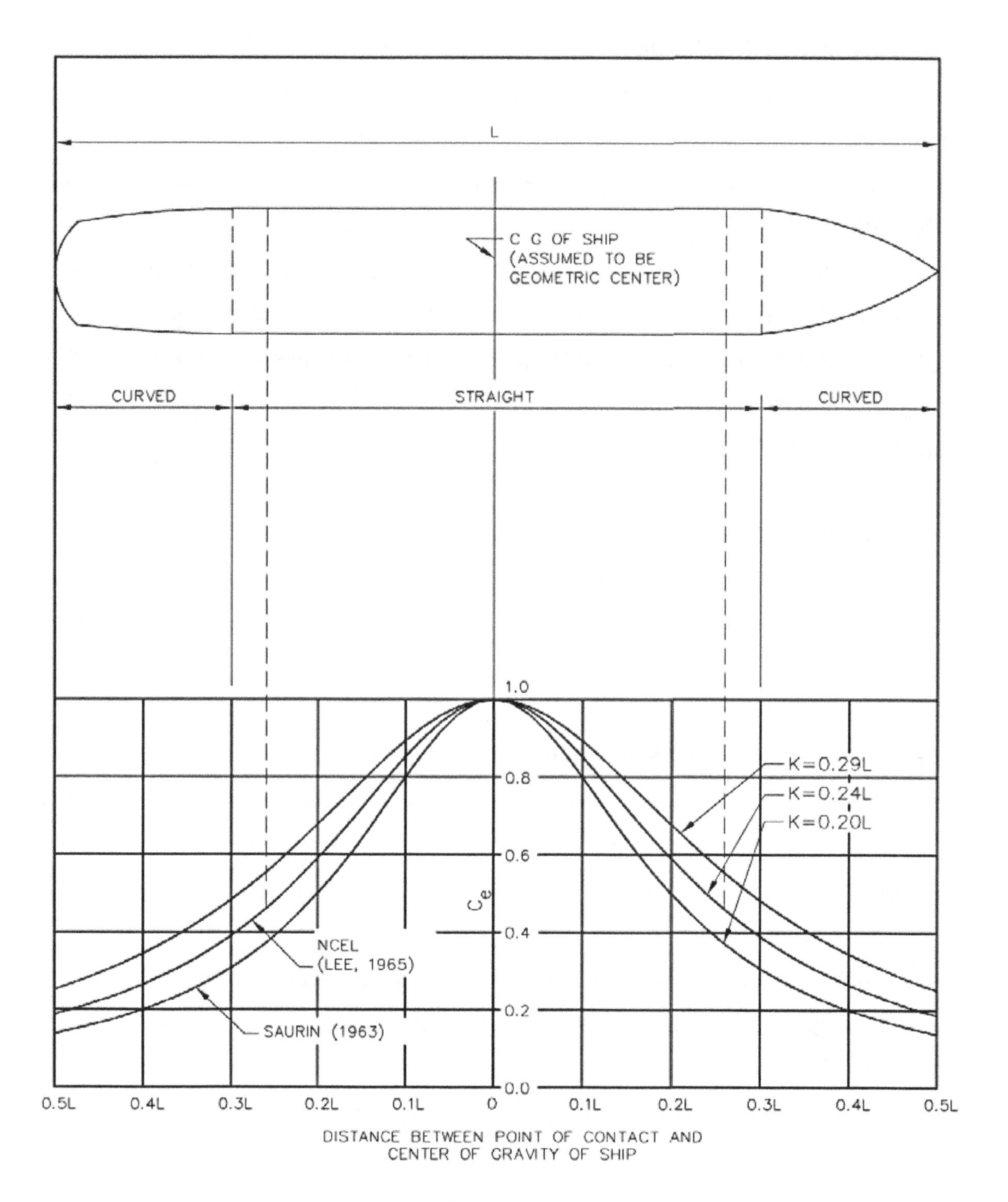

Figure 5-2

Eccentricity coefficient, C_e

2.1.1.6 BERTHING OR APPROACH VELOCITY (V). It should be noted that the kinetic energy of the berthing ship is a function of the square of the normal component of its approach velocity. Thus, the kinetic energy, as well as the resultant force on the berthing structure, is sensitive to changes in approach velocity. By doubling the design value of the approach velocity, the ship's kinetic energy is quadrupled. Design values used for the approach velocity normal to the berth may vary from 0.25 to 1.50 fps (0.076 to 0.46 m/sec), depending on the size of the ship being docked and the tug assistance that is employed. Larger vessels with adequate tugboat assistance can generally berth gently and the lower design velocity may be used. Smaller vessels that self-dock may approach the wharf at considerably higher speeds and, accordingly, the higher design velocity should be used. The berthing velocity is also affected by the difficulty of the approach, maneuvering space for tugs (slip width), and location of the pier or wharf facility. Anticipate higher approach velocities when the berth is located in exposed waters where environmental loads cause difficulty in controlling the ship. Also, currents in tidal estuaries in protected waters can be of major concern. Approach velocity normal to the berth may be taken from Figures 5-3 and 5-4. Determining whether a facility is "exposed," "moderate," or "sheltered" depends on the environmental conditions at the site and is a matter for professional judgment by the designer. Most naval facilities in the United States are situated in protected waters and can be taken as "sheltered." Where high currents (0.3 fps (0.091 m/sec) or more) or strong winds (40 knots (20.4 m/s) or more) occur frequently, a "moderate" condition should be assumed. The "exposed" condition may be used when unusually severe currents and winds are present. However, local experience with ship berthing should control the selection.

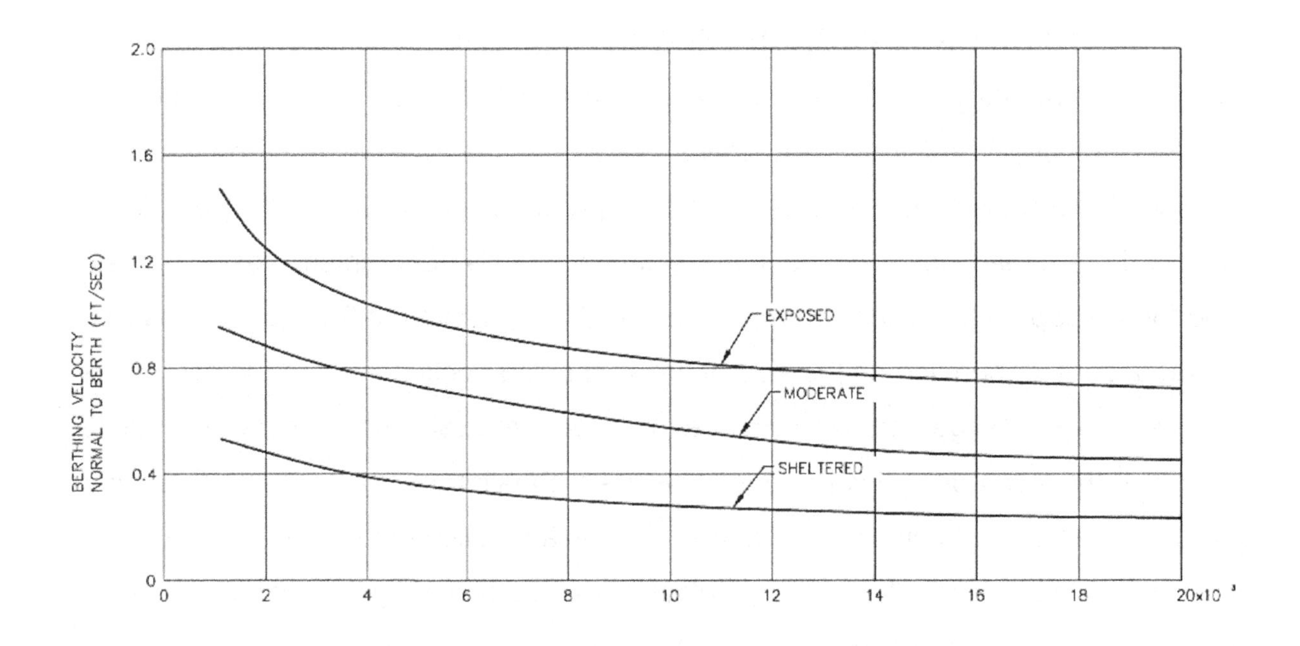

Figure 5-3

Berthing velocity for small ships (up to 20,000 tons)

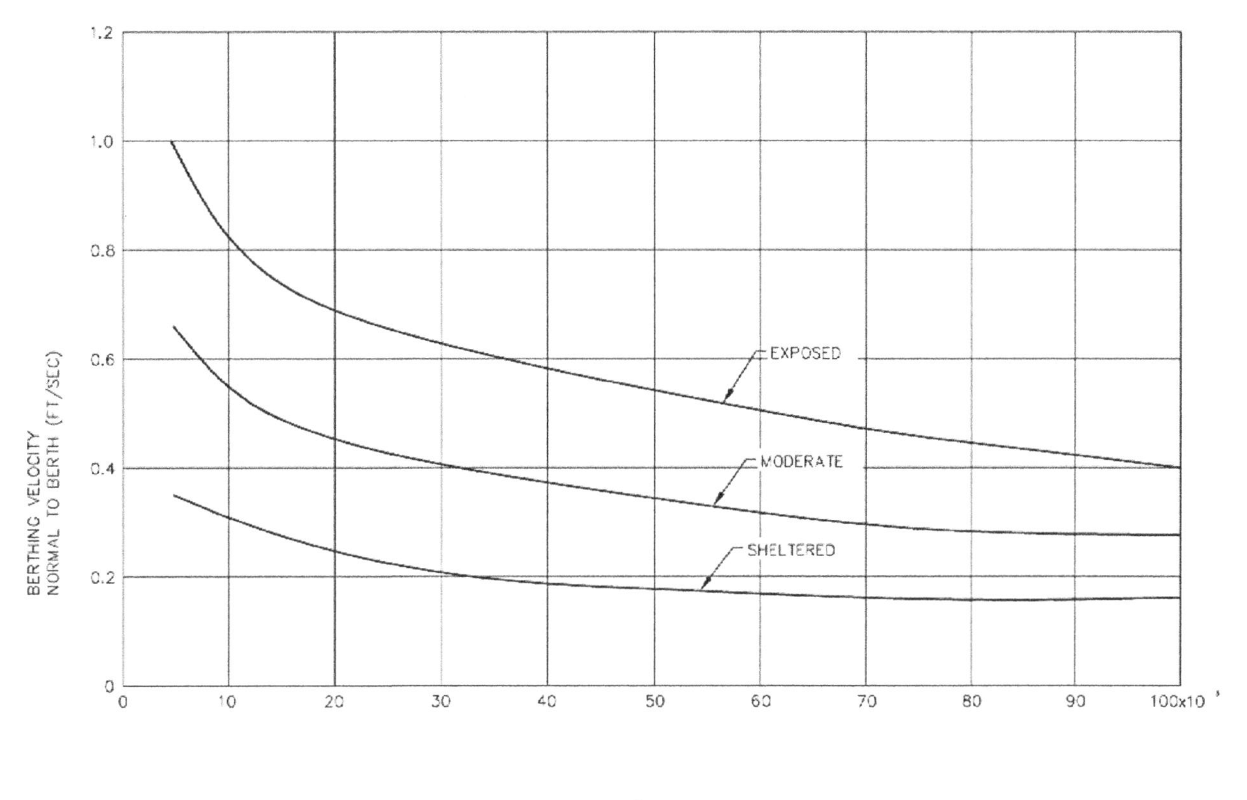

Figure 5-4

Berthing velocity for large ships

2.1.1.7 BERTHING ENERGY FOR SUBMARINES. Procedures for determining berthing energy for submarines are discussed in the technical literature.

2.1.2 BERTHING MODELS.

2.1.2.1 STATISTICAL METHOD. This method is based on actual measurements of the energy of the impact at existing berths. This method is closely related to the conditions of the site where the measurements were taken and is dependent on the fender layout and construction, e.g., distance between piles, and the loading condition of the ship.

2.1.2.2 SCALE MODEL. This method, which makes use of a small-scale model test of the berth to be designed in a well-equipped hydraulic laboratory or ship model basin, suffers from the scale and viscosity effects and requires experienced interpretation.

2.1.2.3 AQWA. Some Owners typically use the COTS AQWA software package to perform a dynamic analysis of berthing through six degrees of freedom in the time domain.

2.1.2.4 ANALYTICAL MODEL. Some Owners developed an analytical model that can accurately predict ship berthing impact forces. The model employs a computational fluid mechanics approach, coupling a Reynolds-Averaged Navier-Stokes (RANS) numerical method with a six-degree-of-freedom motion program for time-domain simulation of ship and fender reactions. The model has been verified with data from small-scale and full-scale tests. Results from the new model for two ship classes yield added mass coefficients close to those calculated by the PIANC formula as long as the depth-to-draft ratio exceeds 1.2. When the underhull clearance becomes small, i.e. for depth-to-draft ratios less than 1.2, predicted added mass coefficients can exceed the PIANC values. Added mass values of 5.0 or higher are predicted for ships berthing against open piers (pile supported) equipped with soft fenders, such as pneumatic or foam-filled cushions. Consider using higher values for the added mass coefficient under these conditions.
2.1.2.5 BeAN. The two previous rigorous analytical models can produce results of superior accuracy, however, they are not always required or always suitable for all cases of analysis. A software program for berthing analysis, BeAN (ref Software Development for Berthing Analysis and Structural Loading on Waterfront Facilities), has been developed. BeAN is a simplified, desktop mathematical model that calculates time histories of fender forces, deflections, and vessel motions. The model examines open or closed type berthing facilities, various depths, eccentric or centric vessel impact, and nonlinear fender structures.

3. TYPES OF FENDER SYSTEMS.

3.1 GENERAL. Fender systems absorb or dissipate the kinetic energy of the berthing ship by converting it into potential energy in the fender materials. This could be in the form of deflection of a fender pile, compression of a column of rubber, deformation of a foam-filled cylinder, torsion of a steel shaft, or pressuring of a pneumatic fender.

3.2 FENDER PILES. This is the most commonly used system in many existing piers and wharves. This system is stiff and lacks the capacity for large deflection which results in high reaction loads and frequent breakage of piling and hull damage. This system employs a series of closely spaced fender piles (5- to 10-ft spacing (1.5 to 3 m)) connected together by chocks and wales. A rubber fender unit is mounted between the wale and the berthing structure. A series of diagonal chains from the structure to the waler completes the system. Tight-fitting joints between chocks, wales, and pile head, with proper tension splices that provide compression and tension continuity along the face of berth must be provided. Ships may be berthed either directly against fender piles or by using additional fenders or camels between the ship and the fender piles. When camels are used, the fender piling must be sized to resist the resulting bending. In the working stress range, there is an approximate linear relationship between reaction force and deflection. When used with floating camels, which tend to cock between the ship and the piles, the ship's energy may become concentrated on just one or two piles. Hence, unless the floating camel is tightly secured to the piles (guided by piles), the system will not work well and frequent damage will occur. The pile-rubber system is not recommended for solid and other types of piers and wharves where full deflection of the piles within the working range will be inhibited. When this system is employed throughout the length of berth, the rubber fender units should be sized for direct berthing of ships (without the use of camels).

3.2.1 TIMBER FENDER PILES. Timber fender piles have historically been the system of choice. Although timber fender piles are still in use, environmental concerns coupled with advances in other fender pile material, have led to a trend in replacement of this type of system.

3.2.2 STEEL FENDER PILES. Steel HP sections, wide flange sections and pipes have been used for fender elements. These pile sections are typically epoxy coated and incorporate a cathodic protection system. Steel fender piles are typically limited to use with foam-filled or hydro-pneumatic fenders since they have a limited range of elastic deflection. The typical steel system is comprised of steel HP sections driven vertically and connected to a steel wale. The wale is attached to the supporting structure by a system of rubber fenders and chains. Steel fender systems have been successfully used. However, due to the additional costs of maintenance and the cathodic protection systems, their use has been somewhat limited when compared to other systems. Plastic coated steel pipe has been used for fender systems in circumstances where large deflections are not anticipated.

3.2.3 CONCRETE FENDER PILES. Concrete fender piles are the most prevalent system in use. Square prestressed concrete fender piles have been tested and proven to have high-energy absorption fendering capabilities and greater strain energy at collapse than either timber or steel fender piling. They are typically 18- to 24-in^2 (1161 to1548 mm^2) prestressed concrete and have served well as primary fender piles for berthing and mooring. Typically, energy-absorbing piles are 18 in^2 (1161 mm^2) and reaction piles (used with foam filled fenders) are 24 in^2 (1548 mm^2) square. The fender piles are usually connected to a chock and wale system at the deck level and supported by rubber fender units at the bullrail. In the working stress range, there is a linear relationship between reaction force and deflection. Figure 5-5 shows a prestressed concrete fender pile installation. Key findings of recent test programs were:

- partial prestressing to 600 psi (4136.9 kPa) was sufficient to close flexural cracks;

- 18 in^2 (1161 mm^2) pile with 20 0.5 in (52 mm) diameter prestress strands in a rectangular configuration confined by No. 3 ties with a 3-in (76 mm) pitch performed best; and
- 65 ft (19.8 m) lengths can be expected to perform well under cyclic load with an ultimate energy capacity of 30 ft-kips (2 m/kN) and a post ultimate capacity of more than 60 ft-kips (4 m/kN.)

Figure 5-5

Foam-filled fender system

(Foam-filled fender, concrete fender piles, steel frame and polyethylene rub strips)

3.2.4 COMPOSITE FENDER PILES. There are two primary types of composite fender piles. One type of composite pile is made of fiber reinforced plastic (FRP) in the form of a tube that can be filled with concrete for greater strength and stiffness. Figure 5-6 shows an FRP fender pile system. Because of a higher susceptibility to abrasion and impact damage, the thermoset FRP tube type pile should have rubberized abrasion strips installed at potential contact points with berthed vessels. The second type of composite pile is made of thermoplastics (such as high-density polyethylene, HDPE) and reinforced with either steel or FRP strands. Figure 5-7 shows a plastic fender pile system. The reinforced thermoplastic type pile generally exhibit larger load-deflection properties compared to conventional wood, steel or concrete piles. To ensure uniform loading and avoid premature failure of fender system components, these type piles should not be used in parallel or mixed with conventional type piles in the same system. Tests have shown that composite piles absorb more energy than timber piles, but are not as stiff. This could pose a problem in the composite fender pile deflects too much. Polymer composite piles are presently at the acceptance phase of development and not yet in the large-scale use phase of development. Much has been happening in the way of codes and standards development for polymer composite piles over the last decade. To date, composite piles have been used primarily for corner protection, as secondary fender piles, and as primary fender piles for small craft facilities.

Figure 5-6

Fiberglass Fender Pile System

(Fiberglass fender pile, plastic pipe wear surface, top support and cap)

Figure 5-7

Plastic fender pile system

(Plastic piles, plastic block and cylindrical rubber fender.

A plastic camel is floating in the water)

3.3 FENDERS.

3.3.1 GENERAL. Some Owners have developed governing specification for selection of marine fenders.

3.3.2 DIRECTLY MOUNTED FENDER UNITS. In this system, individual fender units like the end-loaded, rubber shear, or buckling type, are attached directly to the pier or wharf face. For narrow tidal range in solid piers and wharves, and for narrow vessel size range, this system may be cost-effective for direct berthing of surface ships. Although this system is very popular in commercial piers and wharves throughout the world. This system is subject to damage from snagging on ship protrusions at levels of 8 to 10 ft (2.4 to 3.5 m) above the water line and from vertical loads resulting from snags on rails and protrusions during falling tides or from lateral loads due to snags on protrusion during longitudinal movement of the ship.

3.3.2.1 END-LOADED RUBBER FENDERS. These work by elastic compression of hollow rubber cylinder elements with small length-to-diameter ratios. As shown in Figure 5-8(A), steel fender panels with special rubbing material facing are usually required to minimize wear. The reaction force is an exponential function of the deflection. These fenders are usually attached directly to the pier or wharf structure in the form of a "cell fender."

3.3.2.2 RUBBER SHEAR FENDERS. The potential energy in these units is stored as elastic shear deformation of the rubber. Usually, a solid rubber block is vulcanized between two metal plates and the force is transferred through a fender frame or panel, as shown in Figure 5-9. These fenders are highly sensitive to proper manufacturing and installation as they depend on the bond between steel plates and the rubber. The force-deflection relationship is essentially linear.

3.3.2.3 BUCKLING FENDERS. These fenders operate on the buckling column principle, in which a molded column of rubber is loaded axially until it buckles laterally.

The end-loaded cylinder fenders described earlier are actually a buckling fender in principle. Most buckling fenders are not well suited for direct contact with a moving ship and hence are used with an abrasion or protector panel, as shown in Figure 5-10. The reaction force is linear up to a level when the pure compression behavior changes to the buckling mode. Hence, initially a relatively high reaction is built up with a small deflection, which then stays constant through the rest of the deflection range. Because buckling fenders are intended to buckle in a predetermined direction, any lateral deflection can significantly reduce their effectiveness. When lateral loads in either direction (parallel to length of berth or up/down) are anticipated, a cell-type fender is preferred. These fenders are becoming increasingly popular for berthing very large ships as they can absorb very high energy with a constant reaction force.

3.3.3 SIDE-LOADED RUBBER FENDERS. These are hollow rubber units available in trapezoidal, circular, square, or D-shapes that, when loaded at their side, deform by trying to flatten out. See Figure 5-8(B). The potential energy is stored by a combination of compression and bending of the rubber elements. The reaction force is an exponential function of the deflection and the performance curve is quite similar for all the shapes. Fenders having a curved rather than flat external surface increase in stiffness more gradually as the area of contact increases during deformation. All these fenders experience a sharp and rapid increase in stiffness when the amount of deflection completely collapses the open bore, regardless of their external contour. Side-loaded rubber fenders will not absorb large amounts of energy and generally are not used alone. They are usually provided at the top of fender piles between the wale and berthing structure. A series of diagonal chains from the structure to the wale completes the system. When used with tight fitting joints between chocks, wales, and pile head, and when proper tension splices that provide compression and tension continuity along the face of the berth are used, the system has worked very well in both naval and commercial facilities. Ships may be berthed either directly or a floating camel may be used. When camels are used, size the fender piling to resist the resulting bending.

Both timber and steel piles have been used successfully, with timber being the more common. Prestressed concrete piles are also a viable alternative, exhibiting high-energy absorption capabilities. This type of system provides good berthing flexibility. Ships of different sizes, tug boats, submarines, and barges can be accommodated without any modification. However, when used with floating camels, which tend to cock between the ship and the piles, the ships energy may become concentrated on just one or two piles. Hence, unless the floating camel is tightly secured to the piles (guided by piles), the system may be subject to increased damage. The pile-rubber system is not recommended for solid and other types of piers and wharves where full deflection of the piles within the working range will be prevented. When this system is employed throughout the length of berth, size the rubber fender units for direct berthing of ships (without the use of camels).

3.3.4 FLOATING FENDER UNITS. This system consists of pneumatic or foam-filled fender units and a backing system. As the fender units can be positioned to float with the tide, several surface ship types can be handled. Design the backing system to work with the fender unit for the full tidal range. Because the floating units are usually rather large, they provide a good standoff. The berthing structure may be designed with the backing system at different points along the length of berth and the fender units moved around as berthing plans change. When clustered piles or sheet piles are used for the backing system, additional energy can be absorbed by the piles and their support systems at the deck level.

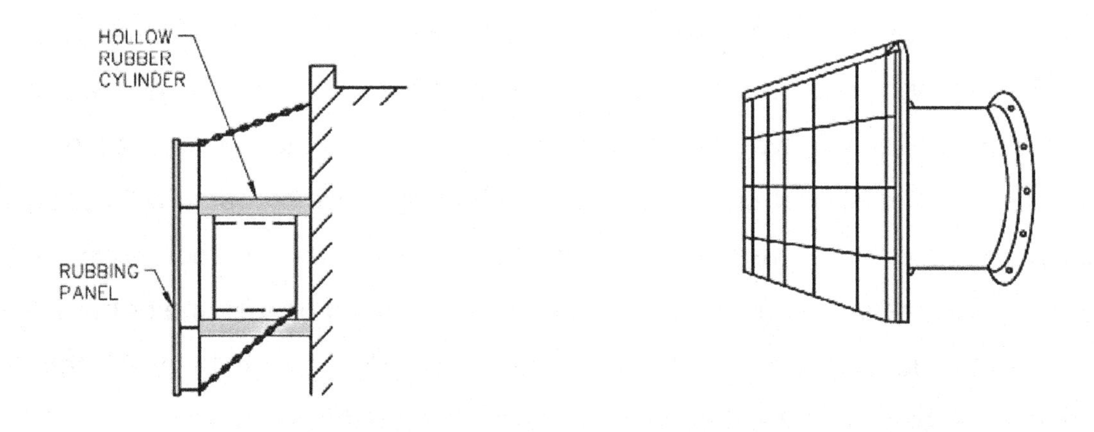

(A) END–LOADED RUBBER FENDERS

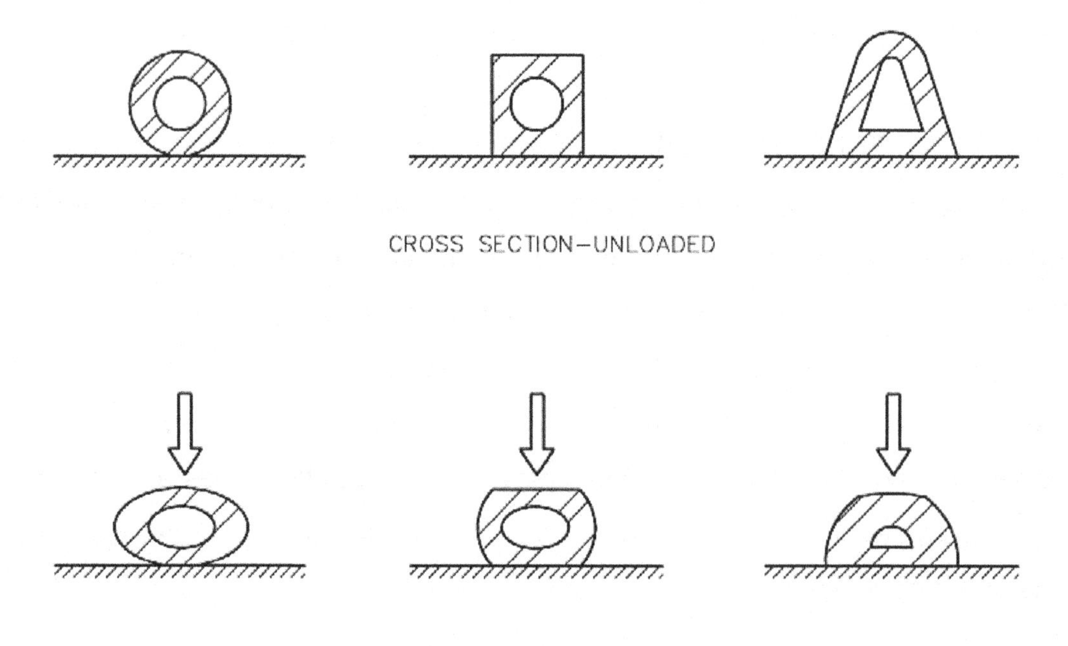

Figure 5-8

Side-loaded and end-loaded rubber fenders

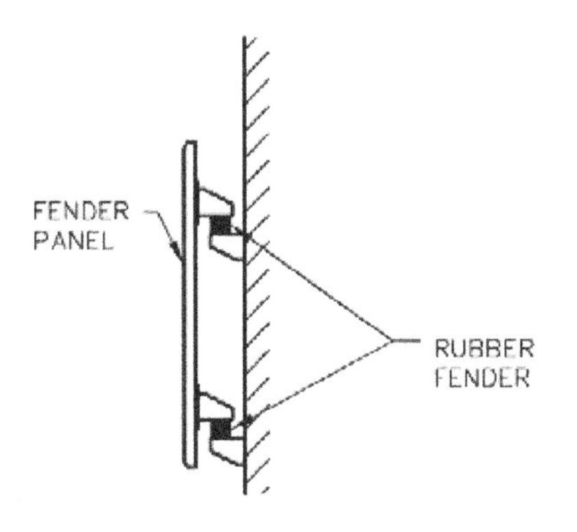

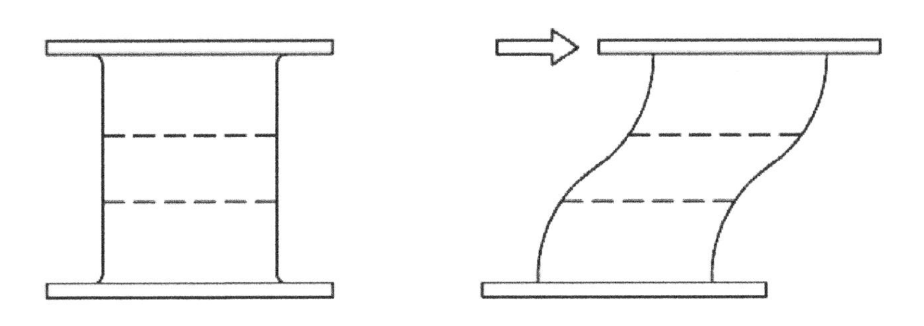

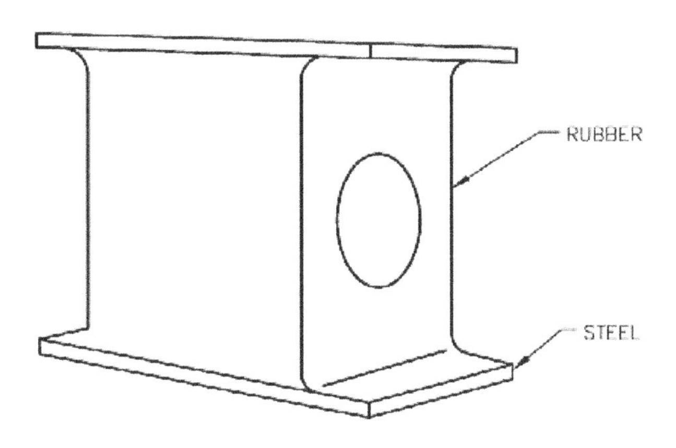

Figure 5-9

Shear fender

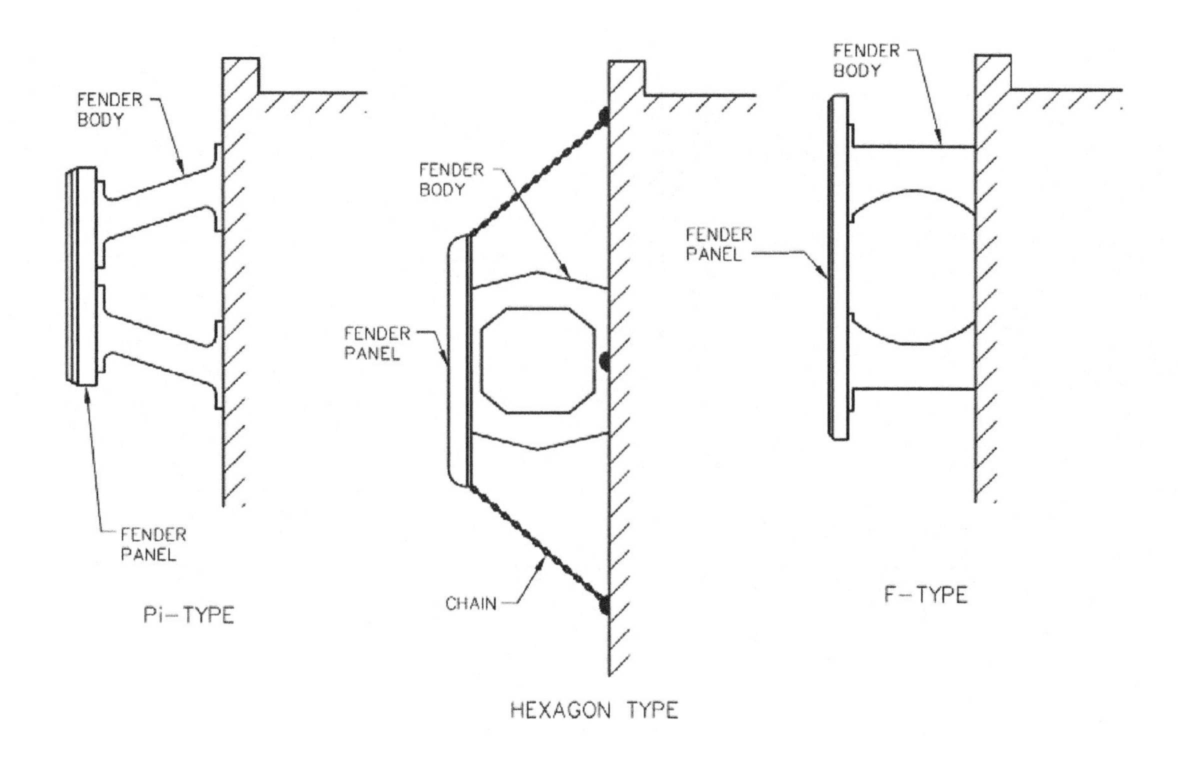

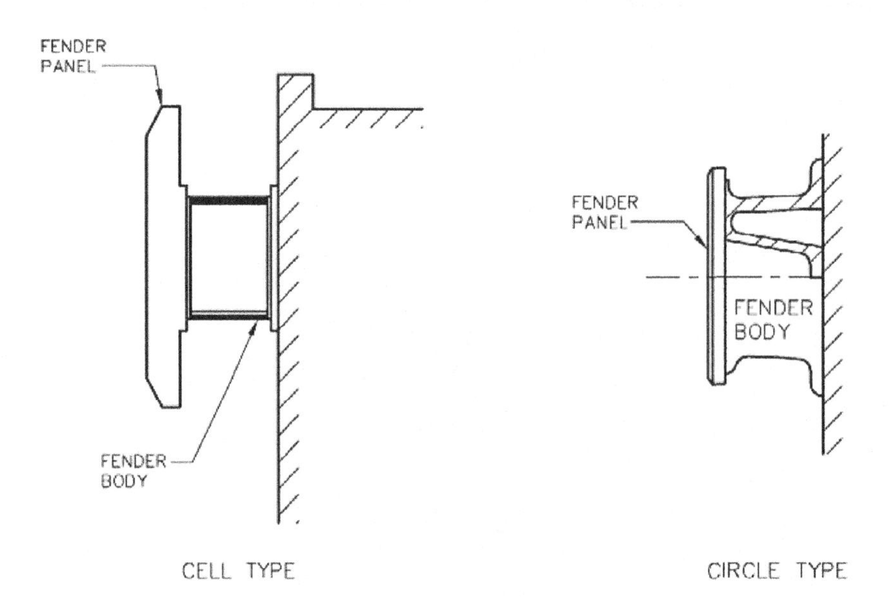

Figure 5-10

Buckling fender with contact panel

3.3.4.1 PNEUMATIC / HYDRO-PNEUMATIC FENDERS. The potential energy in these fenders is stored by the elastic compression of a confined volume of air. By varying the internal pressure of air, the energy-absorption characteristic can be changed. To prevent the air pressure from increasing to a "blowout" level, pneumatic fenders are provided with a relief valve or deflection limiter within the body of the unit. The shell construction for these fenders is similar to an automobile tire with several laminations to provide the high tensile strength required. The surface pressure of these fenders is uniform, resulting in uniform hull pressure. Reaction force is an exponential function of deflection. The basic types of pneumatic fenders in common use are discussed below:

- Air Block and Air Cushion. The shells for these are chemically bonded and mechanically coupled to a rigid mounting plate that can be attached to a solid face of the berthing structure. See Figures 5-11(C) and 5-11(D).
- Floating. The floating type is usually cylindrical in shape with hemispherical ends and is attached to the structure by chains. It floats on the water and rises and falls with the tide. The unit requires a backing system to distribute the load. As shown in Figures 5-11(A) and 5-11(E), large floating pneumatic types are sometimes covered with a net of used automobile tires and cylindrical rubber sleeves to protect the fender from puncture and abrasion. The tire net and chains also form the means for rigging and attaching the fender to the pier.
- Tire. This type consists of a large-diameter tire mounted on an axle and backed by rollers. The unit can be mounted with its axis of rotation vertical or horizontal. This type is particularly suited for pronounced corners of the structure where ships may have approach difficulties. See Figure 5-11(B).
- Hydro-Pneumatic. This type of fender has been developed for use with submarines and consists of a vertically mounted cylindrical pneumatic fender partially filled with water and backed by a closely spaced group of fender piles. A ballast weight is added to adjust the degree of submergence of the fender to coordinate the vertical center of the fender with the horizontal center of the submarine hull; see Figure 5-12. The fender unit floats with the tide and therefore stays in the same relative position with the vessel. For 11-ft (3.3 m) diameter by

32 to 35-ft (9.7 to 10.7 m) high submarine hydro-pneumatic fenders, a vertical and horizontal skin strength of 4,600 lbs/in (82.1 kg/mm) is required. Reduce the hanging counterweight as much as practically feasible. For the fender size mentioned, a hanging counterweight around 13,900 lbs (6305 kg) is recommended.

3.3.4.2 FOAM-FILLED FENDERS. These are constructed of resilient, closed-cell foam surrounded by an elastomeric skin. Provide additional protection against abrasion by thicker elastomeric coatings or an external tire net, similar to the floating pneumatic type. The fender requires a backing system to distribute the load. Netless fenders cost more due to the need for thicker skins and coatings. However, the greater hull marking of the tire net and occasional maintenance need suggest that netless fenders may be preferred. The unit can either be utilized as a floating fender, moving up and down with the tide and held in place with chains or can be suspended from the pier/backing system. The cellular structure of the foam filling reacts like hundreds of millions of individual pneumatic fenders in deforming and absorbing the energy. The foam contains the air within its cellular structure and tends to compress upon itself rather than bulge peripherally. The foam-filled fenders have a high-energy absorption with comparatively small reaction force. Surface pressure of the fender is not quite uniform when it is compressed, so the hull pressure over the contact area is not quite uniform. Where rough concrete surfaces of the backing surface or prestressed concrete piles is a concern, use UHMW pads or strips to protect the skin of the foam filled fender. See Figure 5-5 for a typical foam-filled fender installation.

3.3.5 COMBINATION SYSTEM. Any of the above-mentioned fender systems may be combined in the same berth to make up the deficiency of another. A berth may have either the floating fender units or directly mounted fenders at discrete points, with the in-between spaces filled up by the pile/rubber system designed to work with separators. Floating fenders and directly mounted fenders may be used alternately along solid or double deck types of piers and wharves.

3.3.6 MONOPILE SYSTEM. This fendering system is based on the use of a floating ring-shaped resilient fender unit that rides up and down on a large steel pile driven to the seabed. Special low-friction bearing pads are usually installed on the inner surface of the hull of the ring fender so that the fender unit can rotate and slide freely on the pile. This unique ability makes the monopile system very suitable for corner protection of piers and wharves and entrances to a narrow slip. The units can also work well as breasting and turning dolphins. Energy is absorbed both by the steel monopile in flexure and by the ring-shaped fender unit. This system is illustrated in Figure 5-13.

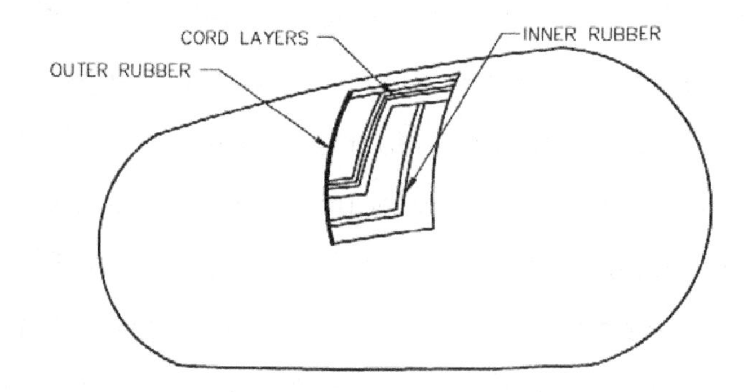

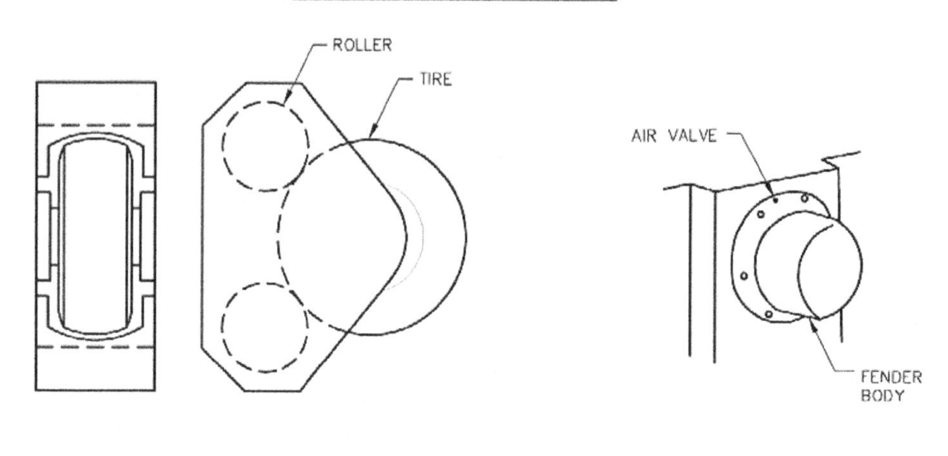

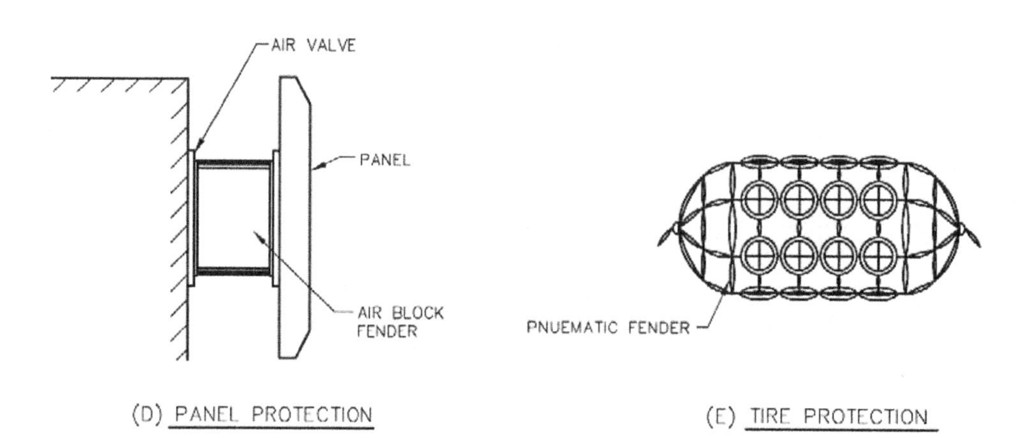

Figure 5-11
Pneumatic fenders

Figure 5-12

Hydro-pneumatic fender and counter weight being installed

4. SELECTION AND DESIGN OF FENDER SYSTEMS.

4.1 GENERAL. The major factors influencing the selection of the best fender system for a particular situation include the following:

4.1.1 ENERGY-ABSORPTION REQUIREMENTS. The fender system must have sufficient energy-absorption capacity to absorb the kinetic energy of the berthing vessel.

4.1.2 REACTION FORCE. This is the force that is exerted on the ship's hull and on the berthing structure during impact. The reaction force has a significant effect on the design of the berthing structure.

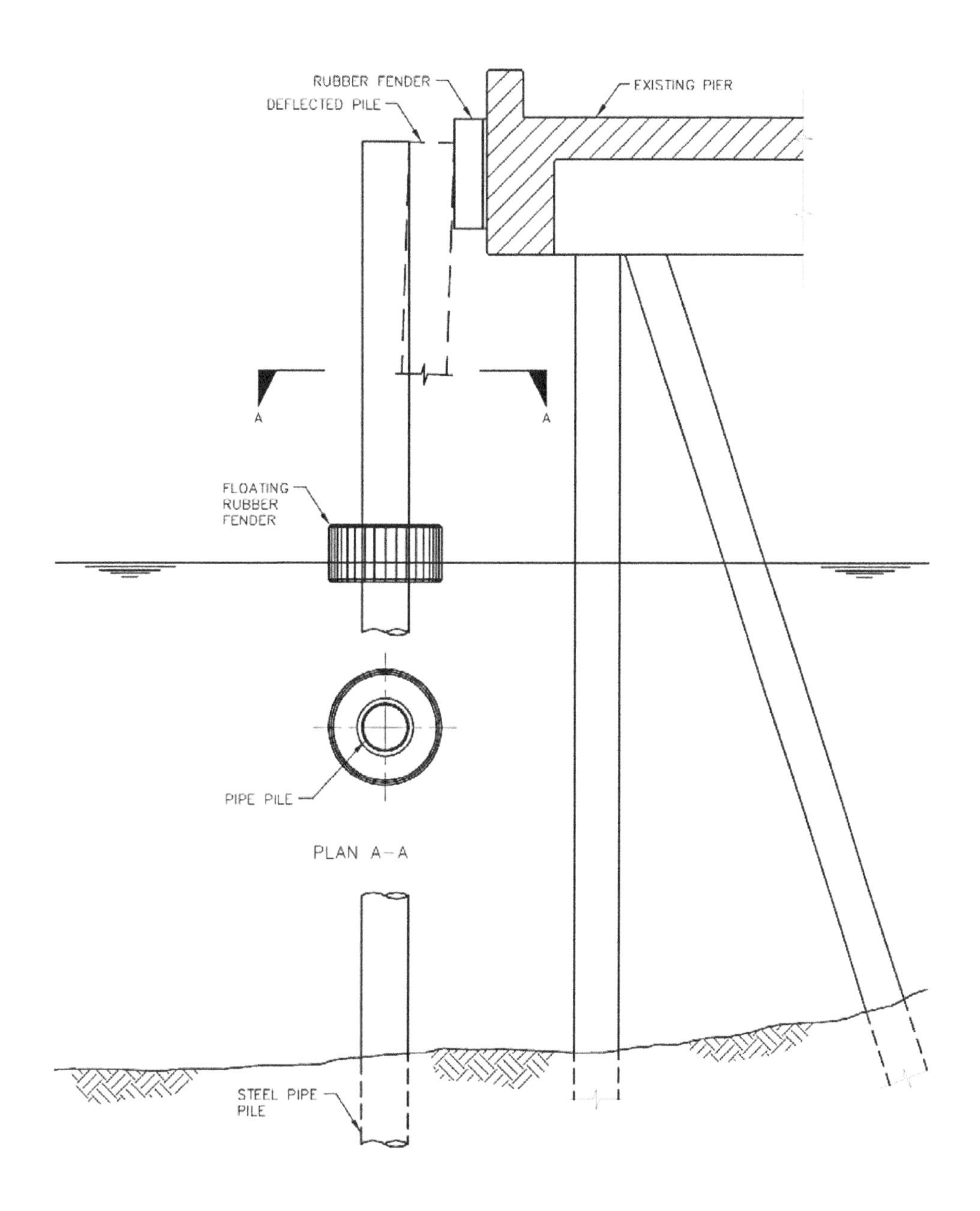

Figure 5-13

Monopile fender system

4.1.3 HULL PRESSURE. This is the pressure exerted on the ship's hull by the fender unit and is derived by dividing the reaction force by the fender area in contact with the ship. Hull pressure must be limited to levels that will not cause permanent damage to the berthing ship.

4.1.4 DEFLECTION. This is the distance, perpendicular to the line of the berth that the face of the fender system moves in absorbing the ship's kinetic energy. The magnitude of the deflection allowable will be controlled by other protrusions from the berthing face and the ship.

4.1.5 REACTION/DEFLECTION RELATIONSHIP. The nature of the reaction deflection relationship determines the relative stiffness of the fender system.

4.1.6 LONG-TERM CONTACT. This includes the changes in environmental conditions (i.e., wind, current, waves, and tide) during loading and unloading at the berth. The fender system should not "roll up," tear, abrade, or be susceptible to other forms of damage when subject to long-term contact.

4.1.7 COEFFICIENT OF FRICTION BETWEEN THE FACE OF THE FENDER SYSTEM AND THE SHIP'S HULL. This determines the resultant shear force when the ship is berthing with longitudinal and/or rolling motion and may have a significant detrimental effect on the energy-absorption performance of the fender system. The magnitude of the shear force also may have a significant effect on the cost of the berthing structure.

4.1.8 DEGREE OF EXPOSURES. Where the berth is exposed to severe wind, current, and/or wave action, the fender selection may be governed by the design mooring conditions rather than berthing conditions.

4.1.9 LIFE-CYCLE COSTS. Evaluate capital costs for both the fender system and the structure; also evaluate costs for operation, maintenance, and repair.

4.1.10 BERTHING PRACTICE. The capability of the crews responsible for berthing the ship will have an effect on the energy-absorption requirement of the fender system. The berthing velocity and angle of approach are affected by the local berthing practice.

4.1.11 MAINTENANCE. Where maintenance is expected to be infrequent, a simple, possibly less efficient, fender system may be preferable to a system requiring a higher degree of maintenance.

4.1.12 LOCAL EXPERIENCE. Consider fender types already used locally because their performance under actual conditions is known. Also, there may be an advantage in having interchangeability of spares, particularly if the number of new fenders required is small.

4.1.13 FREQUENCY OF BERTHING OPERATIONS. A high frequency of berthings normally justifies greater capital expenditures for the fender system.

4.1.14 RANGE OF SHIP SIZES EXPECTED TO USE THE BERTH. While the energy-absorption capacity of the fender system may be selected for the largest ship expected to use the berth, the fender system must be suitable for the full range of ships that the berth will accommodate. The effect of hull pressure and fender stiffness on the smaller vessels may have a significant influence on the selection and arrangement of the fenders.

4.1.15 SHAPE OF SHIP'S HULL IN CONTACT WITH THE FENDER SYSTEM. Where vessels with unusual hull configurations or protrusions are expected to use the berth or where the berth must accommodate barges, pay special attention to the selection and arrangement of the fender system.

4.1.16 RANGE OF WATER LEVEL TO BE ACCOMMODATED. The fender system must be suitable during the full range of water levels that may occur at the berth. The design must consider both the largest and smallest vessels, in both the loaded and light conditions, at high and low water levels. Where extreme water level variations occur, consider using floating fender systems.

4.1.17 CAMELS. The size, type, and number of camels used in berthing operations will seriously influence selection of the fender system.

4.2 FENDER SYSTEM BEHAVIOR. The fender systems having the most promise for new installations can be classified into three groups in terms of their behavior:

4.2.1 FLEXIBLE PILE TYPES. The flexible pile types, or various fender piles discussed earlier, have basically a linear force-deflection relationship. Cantilevered piles or "monopile" systems likewise have a basically linear force-deflection relationship.

4.2.2 BUCKLING COLUMN TYPES. The buckling column types behave linearly up to a point where the rubber starts to buckle and behave nonlinearly from there on. Because the buckling type fender systems have the highest energy-absorption capacity for a given deflection and reaction, they are in very wide use in commercial piers and wharves. Due to the nature of the reaction/deflection/energy-absorption relationship of these types of fenders, a very high reaction (close to maximum) occurs during virtually every berthing operation and the berthing structure must be designed with this fact in mind. This fact also causes the fender to be relatively rigid when smaller ships use a berth designed for larger ones. Many buckling-type fenders cause rather high contact pressures against the ship's hull and consequently require a panel to distribute and thus reduce the pressure. Size and locate the panel to ensure proper contact with both the largest and smallest vessels to use the berth. Another characteristic of these fenders to consider is their lowered performance when impacted by a vessel approaching at an angle to the berth or with a velocity component in the longitudinal direction. The reduction in energy-absorption capacity may be as much as 20 percent when the approach angle is 5 deg. to 10 deg., with additional reduction when combined with shear strain.

4.2.3 PNEUMATIC, FOAM-FILLED AND SIDE-LOADED RUBBER FENDERS. The pneumatic, foam-filled, and side-loaded rubber fenders exhibit very similar behavior with the reaction force building up more than proportional to increasing deflection. The floating pneumatic and foam-filled fenders have a similar appearance and similar reaction/deflection relationship. Compared to the buckling types, these fenders require greater deflection for a given reaction and energy-absorption capacity. The pneumatic and foam-filled fenders present a very large surface to the ship's hull and consequently have low hull contact pressures. This eliminates the need for a panel between the ship and the fender. With the pneumatic and foam-filled types of fenders, the maximum reactions will normally occur only a very few times during the life of the facility, permitting the use of higher stress levels in the supporting structure. However, they require a rather large, solid face on the supporting structure, which may increase its costs. The main difference between pneumatic and foam-filled fenders is that the former will lose its strength completely when punctured by ship protrusions and that the latter may lose a significant part of its energy-absorption capacity under repeated heavy loadings. These behaviors are illustrated in Figure 5-14.

4.3 EVALUATION OF FENDER SYSTEMS.

4.3.1 EQUAL ENERGY AND REACTION. Figure 5-14(A) illustrates the reaction-deflection characteristics of the three types of fender systems. The area under each of the reaction/deflection curves represents the energy absorbed by that type of fender. Each of the curves in the figure represents fender systems with equal rated reactions and equal energy-absorption capability. It is evident from the figure that, while the fenders of the various types illustrated provide equal energy absorption at equal rated reactions, the energy-absorption capacity is achieved through different deflections, with the buckling type deflecting the least.

4.3.2 EQUAL REACTION AND DEFLECTION. A comparison of the various types of fenders may alternatively be considered on the basis of equal rated reaction and equal deflection, as illustrated in Figure 5-14(B). This situation often occurs when new fender units are installed in conjunction with, and compatible with, an existing fender system. It may also occur when a replacement fender system is installed in an existing facility with cargo transfer equipment of limited reach. It is evident from the figure that the buckling type fenders have considerably more energy-absorption capacity for the same reaction and deflection than the other types.

4.3.3 REACTION VERSUS ENERGY ABSORBED. Comparing the various types of fender systems from the point of view of the reaction force that is developed for a given energy-absorption capacity, as illustrated in Figure 5-14(C), it is evident that the pneumatic, foam-filled, and side-loaded rubber type fender units are the "softest." They have greater energy-absorption capacity at reaction levels less than their maximum rated reaction. This characteristic makes these fenders particularly attractive at berths that must accommodate a wide range of vessel sizes because the fenders will deflect significantly even when subjected to relatively small berthing impacts.

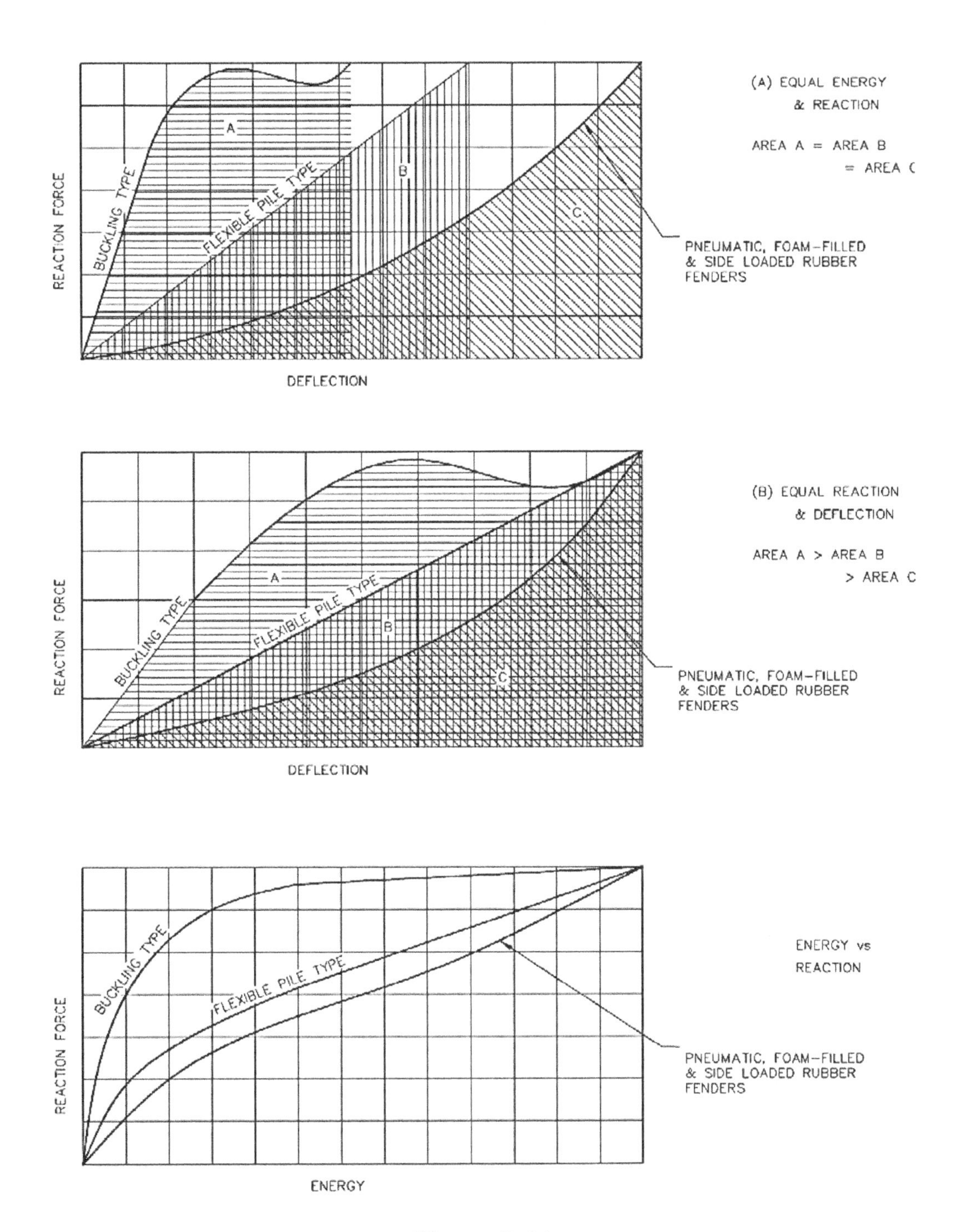

Figure 5-14

Evaluation of fender system types

4.3.4 ACCIDENTAL OVERLOADS. Also to be considered in the selection of a fender system is the consequences of an accidental overload of the system. The buckling and side-loaded rubber fenders "bottom out" if compressed beyond their maximum rated deflection, with resultant high reaction forces and the potential for severe damage to the berthing vessel and the support structure. The reaction of flexible pile fender systems will continue to increase at a uniform rate when overloaded until the yield stress of the pile material is reached, at which point continued deflection will occur as the material yields with no appreciable increase in reaction. Foam-filled fenders, when compressed beyond their maximum rated deflection, will exhibit a steadily increasing reaction and will incur permanent deformation and consequent loss of future energy-absorption capacity. The pneumatic fenders are normally fitted with relief valves so that when overloaded they continue to absorb energy with no increase in reaction beyond that which corresponds to the relief valve setting and no permanent damage to the fender unit.

4.4 FENDER SYSTEM DESIGN.

4.4.1 SHIP CONTACT. While the ideal berthing process would attempt to engage as many fender units as possible, in reality, at the time of impact, the ship will be at a slight angle to the berth and contact will be made over a small length. Design discrete fender units such as the buckling column type or the floating type, for one unit providing the full energy with a minimum of two units installed per berth. For the continuous system using flexible piles and fenders, the length of contact will be a function of the ship's hull radius at the level where contact is made, the flexibility and spacing of fender units, and the stiffness of the chock and wale assembly in the horizontal plane. The problem is analogous to a beam on an elastic foundation. In the absence of more rigorous analysis, the following assumptions for contact length may be made for the following naval vessels:

- 20 ft (6.1 m) for cruisers, destroyers, and frigates.
- 40 ft (12.2 m) for battleships, amphibious warfare ships, and auxiliary ships.

When berthing is made with camels, assume only one camel to be in contact at the time of impact, with a minimum of two camels installed per berth. Where the camels are guided by fender piles, assume all the piles to be effective in sharing the energy. When free-floating camels are used, not all the piles backing the separator will be effective. Local experience should dictate and a more conservative assumption should be made.

4.4.2 ALLOWABLE HULL PRESSURE. When the ship's energy is resisted through foam-filled or pneumatic fenders, the resulting force is concentrated in a small area of the ship's hull. In such cases, the allowable pressure on the ship's hull becomes a critical design issue. Many ships have a thin hull plating with a low allowable hull pressure. Consequently, when checking for an accidental condition, the allowable value may be increased by up to 50 percent.

4.4.3 ALLOWABLE STRESSES. Because ship berthing is a short-term impact type of loading, the following increases over previously published values are permitted.

4.4.3.1 TIMBER. For operating condition, the allowable stress in flexure (tension and compression) may be taken as 0.67 X modulus of rupture or the published allowable values increased by a factor of 2.0, whichever is less. For the accidental condition, the stress-strain curve may be assumed to be linear up to 0.9 X modulus of rupture, which should be taken as the limit.

4.4.3.2 STEEL. For operating condition, the allowable stress in flexure (tension and compression) may be taken as 0.8 X yield stress. For the accidental condition, full yield stress may be used. However, the sections used should satisfy compactness requirements or the allowable stress reduced proportionately. Members should be sufficiently braced for development of the yield strength.

4.4.3.3 CONCRETE. Design reinforced and prestressed concrete members not intended for energy absorption with a load factor of 1.7 over forces developed due to operating condition; they will be satisfactory for the accidental condition. Do not allow further prestressed members to develop tensile stresses in excess of 12 f'c (f'c = 28-day compressive strength) in the precompressed zone. Prestressed concrete members specifically designed for energy absorption will have a maximum allowable working energy (E_{wc}) equal to 0.67 x the applied load (P) x the deflection at the point of the applied load. Three criteria must be satisfied to determine the maximum allowable working energy (E_{wc}) in the pile section for a given length:

- Maximum concrete compressive strain must be equal to or less than 0.0021 in/in (00021 mm/mm.)
- Maximum stress in the prestressing steel at working energy must be equal to or less than 210,000 psi (1 447 898.9 kPa.)
- The working energy (E_{wc}) must be no more than 2/3 of the nominal pile energy (E_{nc}). This provides a minimum factor of safety of 1.5 in case of overload of the pile.

4.4.4 COEFFICIENT OF FRICTION. As the ship is berthed against the fender system, there will be force components developed in the longitudinal and vertical directions also. As the coefficient of friction between rubber and steel is very high, special fender front panels have been developed with reduced friction coefficient. Ultra high molecular weight (UHMW) plastic rubbing strips have been successfully used in front of timber piles. The following friction coefficients may be used in the design of fender systems.

- Timber to steel............... 0.4 to 0.6
- Urethane to steel........... 0.4 to 0.6
- Steel to steel.................. 0.25
- Rubber to steel.............. 0.6 to 0.7
- UHMW to steel.............. 0.1 to 0.2

4.4.5 TEMPERATURE EFFECTS. Fender piles, backing members, etc., are not affected by temperature fluctuations and can be expected to perform normally. However, in colder temperatures, rubber fender units become stiffer and their performance will be affected significantly. Hence, evaluate the energy-absorbing capability of the rubber unit and the fender system as a whole based on the lowest operating temperature expected. Carefully design and detail UHMW rubbing strips which have a significantly higher rate of expansion than steel or concrete to operate effectively.

4.5 CORNER PROTECTION. Provide all corners of piers and wharves and entrances to slips with fender piles, rubbing strips, and rubber fenders for accidental contact with ships or routine contact with tugs. Use any of the different types of fender piles mentioned. See Figure 5-15 for typical details.

4.6 SUPPORT CHAINS. Chains are commonly used in fender systems when a tension member is needed. Chains are used in continuous fender systems and large buckling and cell type units to resist the sudden energy released. For pneumatic and foam-filled resilient fender units, the chain is used to suspend the units. Chain smaller than 3/8 in (9.5 mm) is not recommended. For better corrosion resistance, zinc coating is preferred. A common weldless high-test chain is usually more cost-effective than the stud link variety.

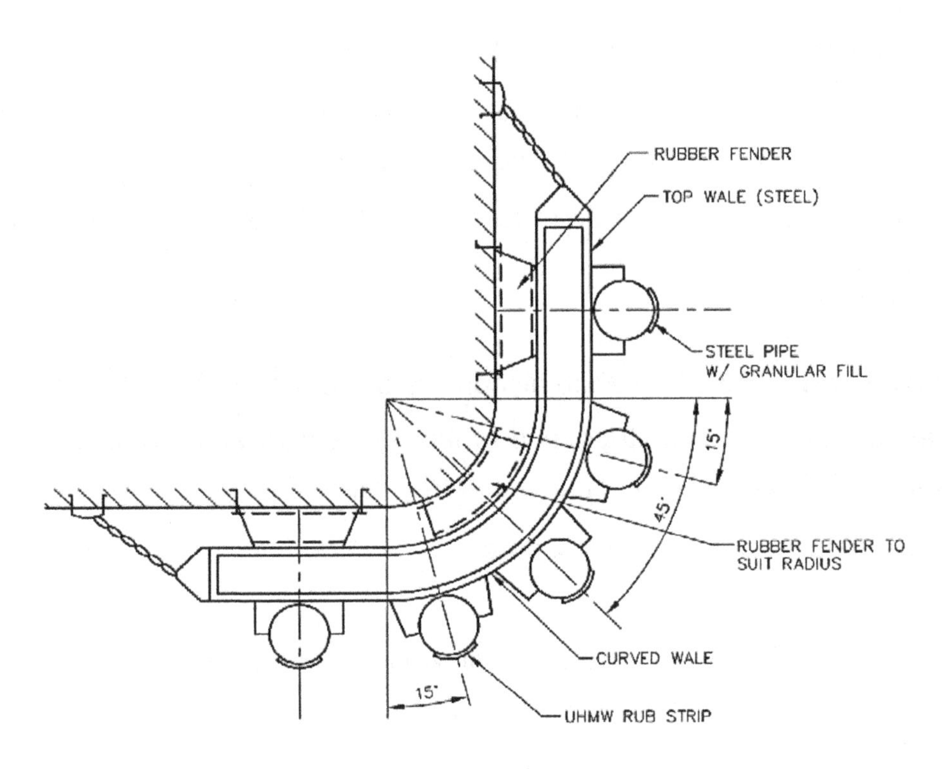

(A) STEEL PILES–ROUNDED CORNER

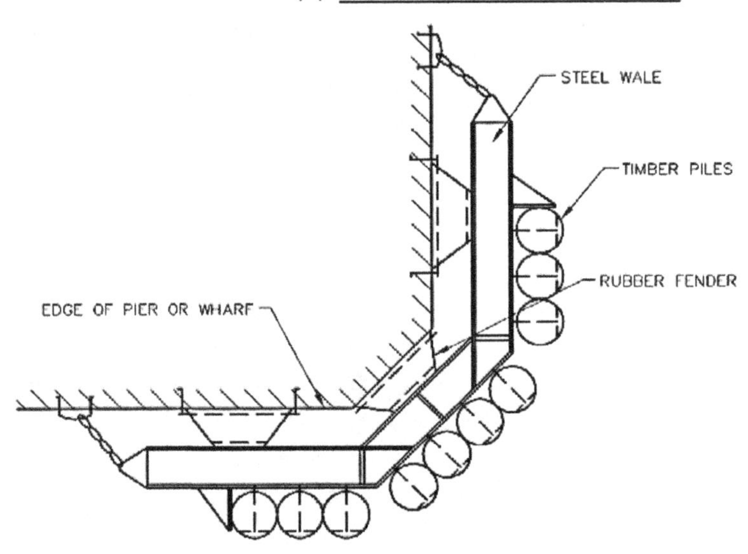

(B) TIMBER PILES–CHAMFERED CORNER

Figure 5-15

Corner protection

CHAPTER 5
CAMELS, SEPARATORS AND ACCESS

1. FUNCTION AND APPLICATION. Camels are devices used between the ship and the pier or wharf structure. Separators are devices used between adjacent multiple berthed (nested) ships to provide a "standoff" or separation. Design of fender piles or other backing system is required to provide the structural interface between the camel and the pier/wharf structure. Camels and separators are used at piers and wharves for the following reasons:

1.1 HULL MAINTENANCE. During active berthing, the ship's crew typically performs cleaning, painting, light hull repairs, and other routine maintenance activities on the ship. These activities are best performed when the ship is kept off the structure at discrete points.

1.2 OVERHANGS AND PROJECTIONS. Ship types having bulges and projections at the side, require camels to prevent damage to the ship at these projections. Other protrusions include air masker bands, soft sonar domes, and stabilizer fins.

1.3 SPECIAL HULL TREATMENTS. Some ships are equipped with special hull treatments that can get damaged through constant rubbing against the structure. Camels and separators with rub strips can minimize the contact area and control the damage.

1.4 SUBMARINE BERTHING. Submarines are typically berthed using camels. The submarines may be moored to the camels or moored directly to the berthing structure. Submarines require the camels to prevent damage to diving planes, screws, fairings, and the special hull treatments.

1.5 MULTIPLE BERTHING (NESTING). Separators are required between ships that have to be berthed abreast for ship-to-ship transfer operations or for lack of berthing space.

1.6 FENDER PROTECTION. When the existing fender system can suffer damage due to motions of moored ships, a camel can be useful in reducing the damage as long as it is properly placed and the ship is properly moored.

2. CAMELS.

2.1 LOG CAMELS. These are large-diameter turned timber logs (24- to 36-in (610 to 914 mm) diameter 40 to 50 ft (12.2 to 15.2 m) long,) held in the desired position from the deck by nylon ropes or chains. They are usually allowed to float with the tide. The longer length is preferred as they can distribute the load to a greater number of piles. Multiple log camels are made from several smaller diameter logs held together by wire rope at ends and at center. They are not as efficient as the single log camels. Plastic coated steel pipes have also been used successfully as camels. The pipe core is generally filled with foam to ensure positive flotation. Log camels fabricated from recycled plastics and composite materials are available. Sometimes a series of used tires may be fitted through the log to provide some energy absorption. Log camels do not provide much of a separation. When a wider separation is needed, other types of camels are more appropriate.

2.2 TIMBER CAMELS. These consist of several large timbers connected together by struts and cross braces to form a large crib. Additional flotation units may be inserted between the timbers for a higher freeboard. Wear causes bolt heads to become exposed and thus cause damage to hulls.

2.3 STEEL PONTOON CAMELS. These are made of cubical or cylindrical steel pontoons connected by structural framing. The pontoons should preferably be filled with foam to reduce the risk of losing flotation by accidental puncturing of the units.

2.4 DEEP-DRAFT CAMELS. For submarine berthing where a good portion of the body is below the waterline, all the above camels are inadequate, as the camel will ride up on the submarine during berthing. Hence, deep-draft camels have been developed. They have limited energy absorption and a narrow working platform. These camels work well when mooring against a tender ship and for multiple berthing. When used against an open pier or wharf, these camels will require solid backing elements (below waterline) from the fender system.

2.5 HYDRO-PNEUMATIC FENDERS. This type of fender has been used as a camel for submarines at a number of locations.

2.6 COMPOSITE CAMELS. These camels are manufactured from composite materials intended to provide an extended service life over steel construction.

3. LOADS. The camel loads are computed from berthing and mooring analysis of the ship, camel, fender, and structure system resisting the lateral loads. All the ship's berthing reaction loads as well as current and wind loads are transmitted through the camels to the pier or wharf structure. Assume all horizontal loads to be acting uniformly along the length. Design deck elements of large camels for 50-psf (2394.0 Pa) vertical live load. Check the camel assembly for fabricated camels for lifting stresses. Where the pick-up points and rigging configurations are critical to control lifting stresses in the camels, provide clearly marked pick-up points or pad-eyes. For complex lifting requirements, provide lifting diagrams on the design drawings.

Figure 6-1

Composite Camel

4. GEOMETRY. The shape and size of fabricated camels are governed by the ship's geometry at waterline.

4.1 SHIP'S LINES. For large vessels, design camels for berthing a single class of ship. Shape the bearing face between camel and vessel to approximate the lines of the ship at waterline. Because hull lines vary with conditions of draft and trim, generally only a rough match is possible. Therefore, provide the outboard camel bearing face with rubber fenders or other means to produce some flexibility in the bearing face, thus compensating for minor hull line variations. Where hull line variations are large, adapters or telescoping devices may be required. Except for straight-sided ships, usually a single line bearing between separators and ships is provided.

4.2 LENGTH AND WIDTH. Provide adequate length of camels in order to keep the contact pressure between camel and hull and between camel and pier fenders within allowable limits. This is particularly important where compressible fender faces are used that transfer reaction pressure directly to the hull plate versus the frames of the vessel. The length should not be less than the distance between three frames of the ship, three fenders or fender piles on the pier, or 30 ft (9.1 m,) whichever is greatest. Minimum camel width is determined by the ship's roll characteristics and freeboard, the presence of any overhanging projections on a ship and vertical obstructions on the dock such as gantry cranes or light poles.

4.3 DEPTH. Provide adequate depth for submarine camels and separators to maintain contact with ship and pier/wharf in the full tidal range.

5. STABILITY. There is usually some eccentricity between the horizontal load applied on the ship-side and the horizontal reaction provided on the dock-side. This is due to tilting of the camels (from imperfect flotation, buoyancy tank taking on water, etc.) and tendency of the camel to ride up and down with the vessel due to tidal fluctuations. The camel should have sufficient width, depth, and weight to provide roll stability for counteracting the effect of the load eccentricity and should have means of adjusting for variations in tilt and trim.

6. LOCATION. For fine-lined ships, camels should generally be placed within quarter points of the ship to give strength and to bear on a reasonably straight portion of the hull. For straight-sided ships, camels may be located beyond the quarter points. Provide at least two camels for each class of ships. Camels should not be placed so as to bear directly against structural piles.

7. MISCELLANEOUS CONSIDERATIONS.

7.1 PROTECTION. Consider outfitting steel separators with a suitable protective coating or a cathodic protection system, depending on relative costs.

7.2 BUOYANCY TANKS. Buoyancy tanks should be compartmented or foam filled. Provide drainage plugs used for pressure testing the buoyancy tanks. Where pontoon camels are assembled in a single line, they should be ballasted for stability through plugged openings provided for this purpose. Consider the buoyancy of framing members and the weight of paint, if any, in the buoyancy and stability computations. Where buoyancy tanks are not foam filled to allow filling with ballast water or weights to adjust trim and freeboard of the camel, provide easily accessible fill/pumpout and vent connections. These connections can be used to pump out excess water that leaks into the tanks.

7.3 ABRASION. Camel fenders rubbing against a hull remove its paint. Exposed surfaces are subject to corrosive action, especially at the waterline. For these reasons, it is desirable to have camel fenders rub against hulls above the waterline where the hull can be repainted if necessary.

8. CAMEL DESIGNS.

8.1 SUBMARINE CAMELS. Tapered and non-tapered camels are designed for use with specific submarine classes.

8.2 AIRCRAFT CARRIER CAMELS. Camels are usually proprietary designs.

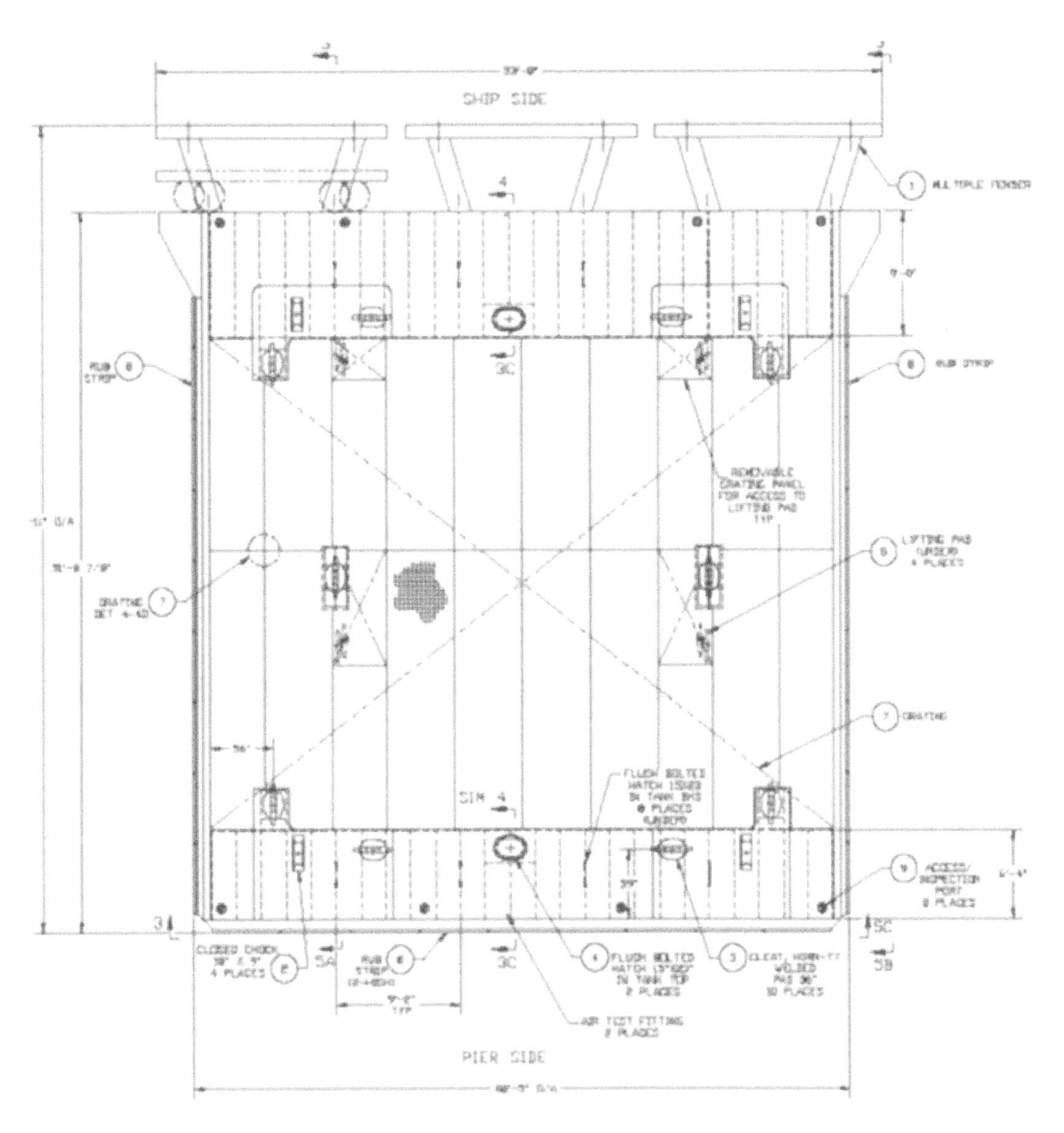

Figure 6-2

Aircraft carrier camel

The following pertinent information is provided for a representative aircraft carrier camel:

- Construction – steel floating frame pontoon
- Loads - 1500 kip (6672.3 kN) lateral and 500 kip (2224.1 kN) longitudinal reaction force transmitted to the camel fender system along the length of the camel. Values were based on maximum wind and current forces associated with heavy weather conditions (95 mph (42.5 m/s) wind and 1 knot (0.51 m/s) current) and breasting reaction associated with CVN berthing assuming a maximum approach velocity of 0.5 ft/sec
- Suitable for placement along the flat side of the hull approximately between frames 110 and 190 with an allowable hull pressure of 30 psi (206.8 kPa.)
- Working surface designed for 100 psf (4788.0 Pa)
- Size: 48'-5" (14.7 m) wide with standoff of 57'-11" (17.7 m) (undeflected fender)
- Weight: 95.4 long tons
- Light draft: 3.66' (1.1 m); Live load draft: 4.34' (10.4 m)
- Light freeboard: 3.3' (1 m); Live load freeboard: 2.4' (0.73 m)

9. SEPARATORS. Separators are usually designed for a specific class of ship.

10. ACCESS

10.1 GENERAL. Several access structures are used in piers and wharves for moving personnel and cargo, and accommodating selected utility lines. They have unique design requirements. Some are of standard design while others are designed and constructed specifically to go with the facility. The access facilities covered by this discussion are landing float, brow or gangway, brow platform, walkway or catwalk, and ramp. Where there is a potential need for access by a physically handicapped person, meet the applicable requirements of the Americans with Disabilities Act Accessibility Guidelines and the Uniform Federal Accessibility Standards, and when they differ, the one with the greatest accessibility requirement will govern.

10.2 SAFETY. Safety on piers and in pier design is an important consideration. Consider operational uses and where possible provide safety precautions, protections and warnings to minimize the potential hazards. Double deck piers require additional safety considerations. Potential uses and needs for pier deck space are unlimited and each must be evaluated based on the local requirements. It is essential that safety and operators be involved in the pier design, so potential uses can be identified, categorized and policy determined. In particular high-voltage electrical, fall hazards, trip hazards, and access to industrial or mechanical areas are issues that require additional consideration. In light of this, a study was conducted in which numerous activities were examined in detail and a matrix developed specifying essential use activities as well as prohibited activities. The study provides useful information such as details on all signage required and detailed plan views of both upper and lower decks for double deck piers which show:

- Utility locations
- Activity areas
- Striping
- Transit paths
- Vehicle turn locations

- Curb, fencing, guardrail, handrail, lighting, life ring, fall protection and signage locations

10.3 LANDING FLOAT. When piers and wharves need to be accessed from the waterside by small craft such as patrol boats (which cannot berth directly), a landing float and a brow are required.

10.3.1 MATERIALS. Flotation units may consist of foams of polystyrene and polyurethane, fiberglass-reinforced polyester resin shells with or without foam cores, metal pontoons, metal pipes, metal drums, and hollow concrete sections. Timber logs, the earliest form of flotation unit and the cheapest, have a tendency to become waterlogged and their use is not recommended. Decks of floats are variously made out of wood planks, plywood, plywood and fiberglass-resin coatings, concrete, and nonskid metal surfaces. Framework for floats is generally of preservative-treated timber, although steel and aluminum are often used. All ferrous metal hardware should be galvanized or otherwise protected from corrosion.

10.3.2 MOORING SYSTEMS. Anchor floats to prevent movement by wind, current, waves and impact from the ships. Anchorage may consist of individual vertical (guide) piles, frames of batter and vertical piles, and cables or chains. When piles are used to anchor small floats, guides are furnished to secure the float to the anchor pile. Commonly used guides are rigidly braced metal hoops of pipes, rollers, or traveler irons. Chains and flat bar guides should not be used as they cause the float to hang up on the piles. See Figure 7-1 for details. This system works well for shallow waters with a large tidal range. In deeper water, the pile head may have to be supported by the structure or pile driven deeper. Anchorage may also be obtained from a cable or chain system attached to the ocean bottom or to the fixed pier or wharf structure.

10.3.3 LIVE LOADS. Design stages for landing personnel only for a uniform live load of 50 lbs/ft2 (244 kg/m2) or a concentrated live load of 500 lbs (226.8 kg) placed at any point on the deck surface. The float should not tilt more than 6 deg. from the horizontal when applying the concentrated live load of 500 lbs (226.8 kg.)

10.3.4 FREEBOARD. Floating stages for small craft usually ride with the deck from 15 to 20 in (381 to 508 mm) above the water surface under dead load. Live loads usually lower the float about 8 to 10 in (203 to 254 mm).

10.3.5 FENDERING. Provide fenders on all floating stages. For small craft berthing, fenders may consist of soft, flexible rubbing strips (rubber tires, sections of hose).

10.3.6 FITTINGS. Provide a minimum of three cleats (5,000-lb. (2268 kg) capacity) for securing small craft.

10.3.7 FINISH. Provide the deck with a nonskid surface. Where wheels or rollers from a brow will be resting on the float, provide guide channels or a skid plate to prevent damage to the float.

10.3.8 REINFORCED PLASTIC LANDING FLOAT. The float shown in Figure 7-2 is constructed of a planking material, referred to as "rovon planks," formed by wrapping glass roving, spirally, around rigid polyurethane foam cores. For extra strength, several wrappings may be applied. The float is 60 ft (18.3 m) long, 14 ft (4.3 m) wide, and 5 ft 4 in (7.6 m) deep. It weighs about 26,000 lbs (11793.6 kg.) The deck is covered with a nonskid coating. In unloaded condition, the float draws 2 ft 1 in (0.64 m) of water and in a loaded condition the float, designed for a uniform live load of 100 lbs/ft2 (488 kg/m2) or a concentrated load of 500 lbs (2268 kg) placed at any point on the deck, draws 3 ft 7 in (1.1 m) of water. Cleats and a timber fender system are provided. A 12- by 12-in (305 by 305 mm) timber member is attached at each end to receive timber pile guides located at each corner. For additional details, refer to R605. The float is light, strong, and has a high roll stability due to the catamaran-type hull construction.

10.3.9 CONCRETE FLOAT ELEMENTS. Concrete encased plastic foam elements designed for use in concrete floating docks for marinas can be connected together in various configurations to be used as work floats. The mass added by the concrete encasement creates a very durable float that is less affected by waves and live loads than more lightweight systems.

10.4 BROW OR GANGWAY. Brows are used for access to landing floats from the pier or wharf structure, however they are more frequently used to provide personnel access from a pier or wharf to a berthed ship.

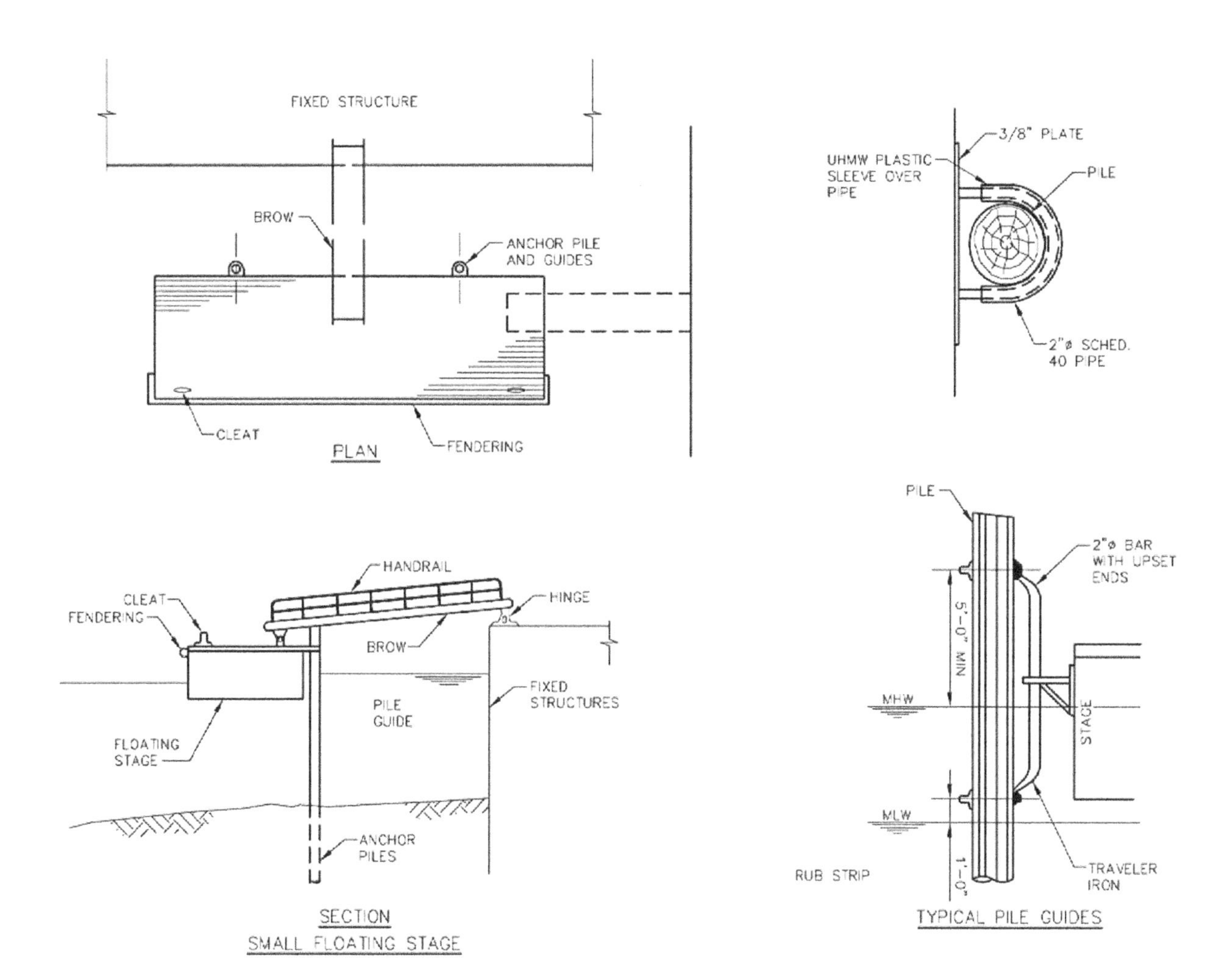

Figure 7-1

Small floating stage

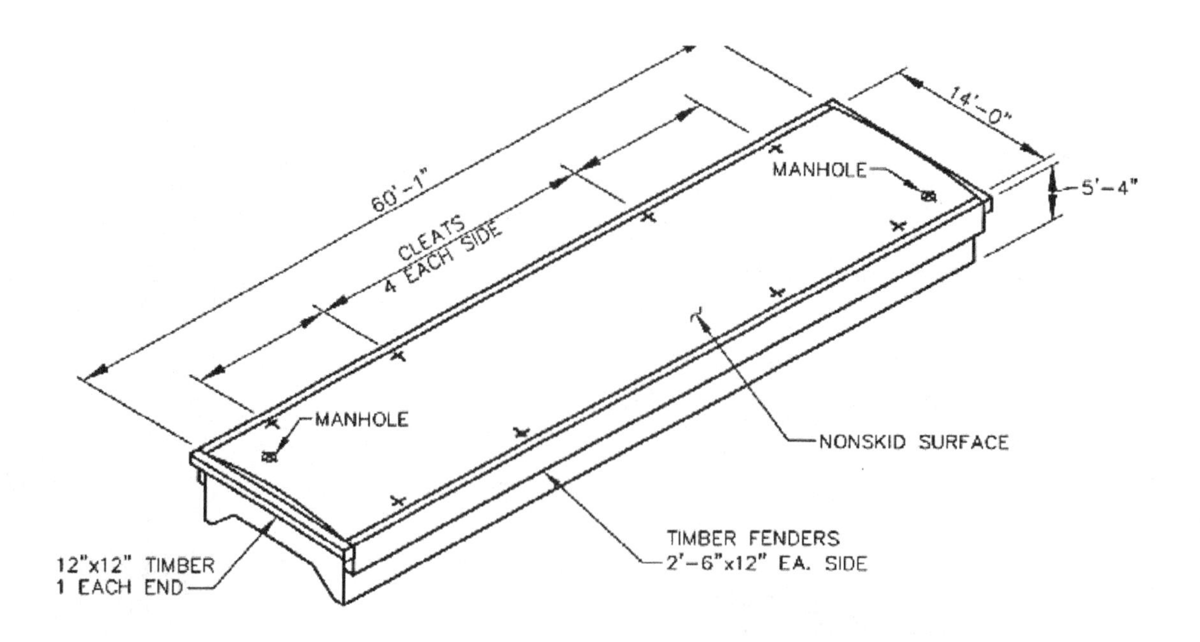

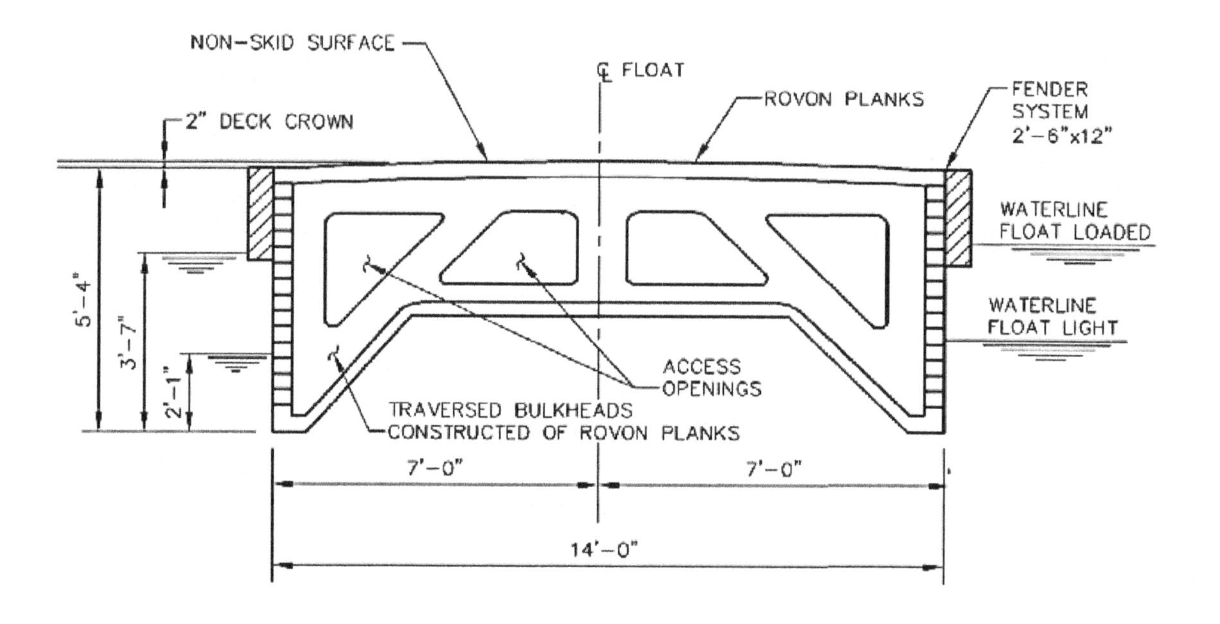

Figure 7-2

Reinforced Plastic Landing Float

10.4.1 LENGTH. Brows should be of sufficient length so that the slope will not exceed 1.5 horizontal to 1.0 vertical at the worst condition.

10.4.2 WIDTHS. Widths should be 36-in (914 mm) minimum (clear) passage for one-way traffic and 48-in (1220 mm) minimum (clear) passage for two-way traffic. Provide a 60-in (150 mm) minimum (clear) passage for two-way traffic when personnel carry small loads.

10.4.3 CONSTRUCTION. Use fiberglass, aluminum, steel, timber, or a combination of these materials. Aluminum and fiberglass are generally preferred for the low weight to strength ratio and corrosion protection.

10.4.4 LIVE LOAD. Design the brow structure for a uniform live load of 75 lbs./ft^2 (366 kg/m^2) and a concentrated live load of 200 lbs (90.7 kg) applied anywhere. A reduction in the live load to 50 lbs./ft^2 (244 kg/m^2) may be permissible where the brow is to be used in conjunction with a landing float. For calculation of reaction to the landing float, the live load can further be reduced to 25 lbs./ft^2 (122 kg/m^2.)

10.4.5 HANDRAILS. Design handrails to resist 50 lbs per lineal foot (75.6 kg per lineal meter,) applied in any direction at the top. Design handrails to resist a single concentrated load of 200 lbs (90.7 kg,) applied at any direction at any point along the top. It is assumed that the uniform load and the concentrated load do not act concurrently. The handrail may be designed to serve as the top chord of a truss when sufficiently braced.

10.4.6 SAFETY. Provide safety devices to keep the brow from rolling off the platform deck and to prevent movement of the platform while in use. Clip safety chains into position for personnel safety. Large tidal variations are a problem because these may cause the brow to roll off the platform. A similar situation exists when high winds, currents, and extreme tides pull a ship away from the pier. Numerous accidents have been associated with brows being supported by pallets either on the ship end or pier end of the brow. Brows are not to be used in conjunction with pallets. It a platform is necessary, it shall be properly engineered, designed, and built.

10.5 BROW PLATFORMS. Brow platforms are used when a brow from ship deck to pier deck is not practical, or presents an obstruction. Examples are portal crane trackage along repair berths, large tidal variations, and great height from deck to pier. Aircraft carriers usually use one brow forward and two aft. These brows require platforms 20 ft (6.1 m) or higher. This platform is basically a truncated tower, with typical measurements of 12 x 12 ft (3.7 by 3.7 m) at the base, while the top deck is 5 ft (1.5 m) wide and 10 ft (3 m) long. If small stair platforms are built alternately opposite hand, the requirement for a large platform can be met by lashing two of the smaller ones together. Sometimes the ship end of the brow can be connected to a rotatable platform which is permanently fixed to the ship by means of pins that lock the brow pivot hooks to the circular rotating portion of the rotatable platform. Construction materials and live load requirements are the same as for brows.

10.6 BROW AND PLATFORM DESIGNS. Standard designs for brows and platforms may be available.

10.7 WALKWAY OR CATWALK. These are permanent personnel access bridges installed between shore and different elements of piers and wharves. One example is a walkway between the pier or wharf structure and a mooring dolphin located some distance away.

10.7.1 WIDTH. For walkways between shore and a U-shaped wharf, a 4-ft (1.2 m) width is recommended. For infrequently used walkways, the minimum width should be 3 ft (914 mm.)

10.7.2 LIVE LOAD. Design all walkway structures for 100 lbs./ft2 (488 kg/m2) live load.

10.7.3 CONSTRUCTION. Walkway decking should be slip-resistant aluminum or fiberglass grating. Framing may be wood, aluminum, or fiberglass members. In view of the light loads encountered, piles supporting deck stringers can be of treated timber. Where loads and installation difficulty make timber piles inadequate, concrete and steel piles may be used.

10.7.4 HANDRAILS. Provide handrails on either side of the walkway. Consider handrails for use along edges of approach trestles and along non-berthing extents of docks or wharves.

10.8 RAMPS. Transfer bridges or ramps are sometimes required for moving vehicles or heavy cargo from ships, similar to a roll-on/roll-off (RO/RO) operation. Sideport ramps are stowed and handled by the ship. Sternport ramps are hinged to the vessel and extend to dockside or floating equipment (lighters, causeways, stages). These ships also have conventional cargo gear. The LHA-class of amphibious assault ships has vertical lift stern gates, possess RO/RO capability. Installations accommodating vessels of this type should consider the use of a ramp or transfer bridge, as shown on Figure 7-3, to minimize the time required for movement of vehicular equipment and for loading of supplies. Design and construction of ramps should be similar to highway bridges. One key issue regarding ramps is clearance from pier/wharf edge obstructions, i.e. utility risers, cleats, bollards, etc. When designing piers and wharves, consider the specific location for ramp landing as well as structural deck strength. Sideport loading ramp access for LHA, LHD, and LPD-17 class ships is an important consideration, particularly on double deck piers.

10.9 ACCESS LADDERS AND LIFE RINGS. Provide ladder access from pier or wharf deck to waters at a maximum spacing of 400 ft (121.9 m) on centers or within 200 ft (61 m) of the work area per 29 CFR 1917.26, First Aid and Lifesaving Facilities. Such ladders should be at least 1 ft 4 in (0.38 m) wide and should reach the lowest water elevation anticipated. Safety cages are not required. Locate the ladder on either side for a pier (50 ft (15.2 m) or more wide) and on the waterside for a wharf at places convenient to anyone who might accidentally fall into the water. Also, 29 CFR 1917.26 requires that a U.S. Coast Guard-approved 30 in (762 mm) life ring with at least 90 ft (27.4 m) of line attached be available at readily accessible points at each waterside work area where the employee's work exposes them to the hazard of drowning. Interpret as one life ring per wharf.

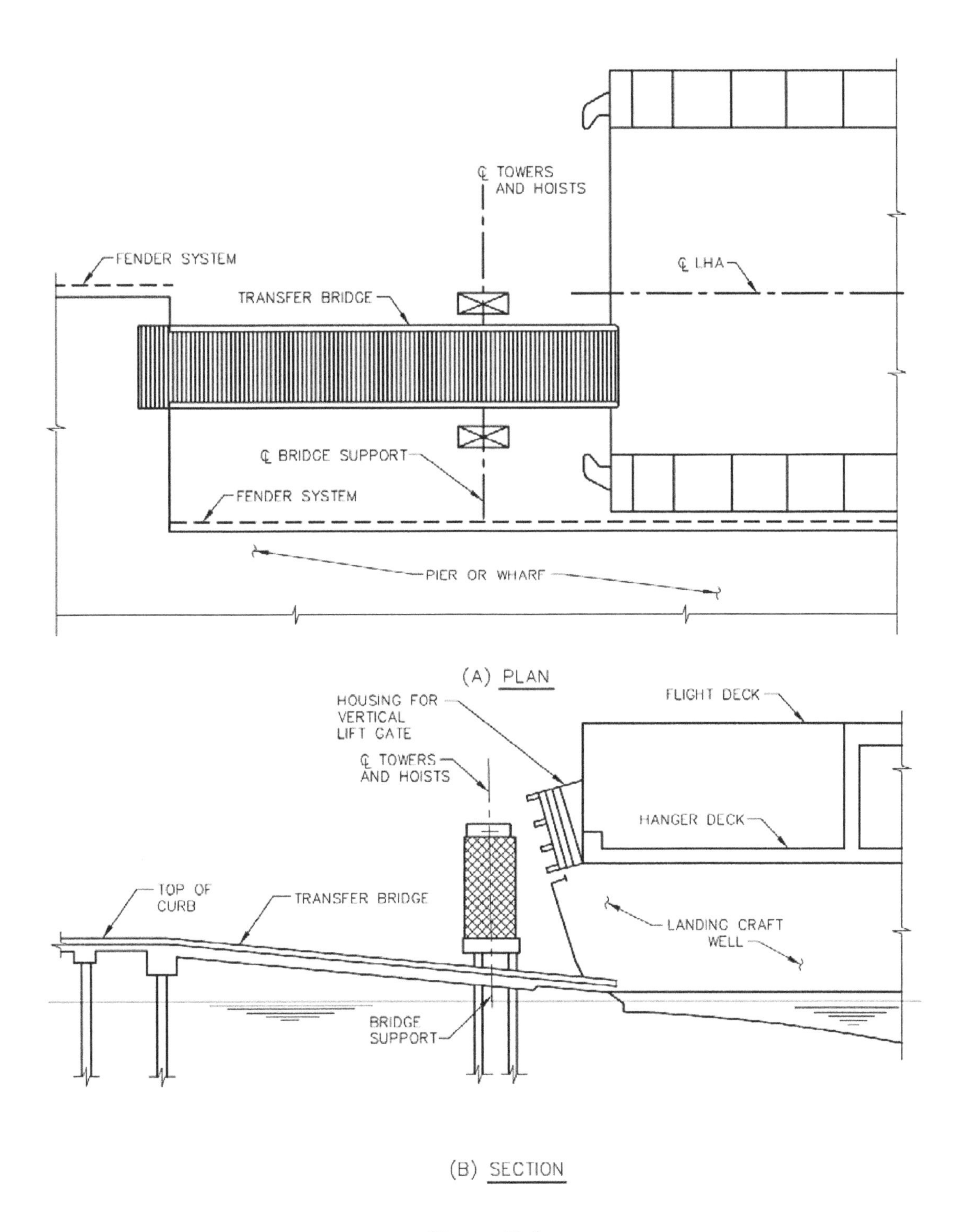

Figure 7-3

Transfer bridge for LHAs

CHAPTER 6
MARINE FUELING FACILITIES

1. FUNCTION. Design marine fuel receiving and dispensing facilities for the purpose of receiving fuel and/or loading fuel aboard ships, barges and boats for consumption or as cargo. In many cases, the marine receiving and dispensing facilities will be combined.

2. FUEL PIERS AND WHARVES. When required and approved by the Owner, design fuel piers for dispensing and receiving fuel. Ensure that the size of the facility is compatible with the fuel requirements of the activity and the number of simultaneous loadings and off-loadings to be accommodated. For dispensing of fuel, consider the number, type, and size of vessels to be fueled or loaded to provide the required number and locations of fuel outlets. In most cases, use dedicated fuel piers and wharves for fuel receipt. Include in the design an energy absorbing fender system.

3. BERTHING PIERS. In some cases, permanent fuel piping and equipment may be installed on berthing piers which were not primarily designed for handling fuel. These facilities are normally used only for dispensing fuel to surface combatants for consumption. Operational requirements usually dictate a clear berthing pier surface area. This imposes restrictions on the use of loading arms and above deck piping. For these areas, trench-contained piping may be considered. Prior to designing facilities on berthing piers for receiving and/or dispensing of bulk fuel for transport, review plans with appropriate port operations agency.

4. OFFSHORE MOORINGS. When operations of an activity do not warrant construction of fuel piers, provide offshore moorings for vessels to discharge or receive fuel through underwater pipelines connecting to the shore facility. Clearly mark the moorings so that the vessel, when moored, will be in the proper position to pick up and connect to the underwater connection.

5. GENERAL REQUIREMENTS. Do not start the design of any fueling system without first becoming completely familiar with guidance on spill prevention, air quality control, and other environmental, safety and fire protection issues.

6. GENERAL LAYOUT. Provide pier loading and off-loading connections, with blind flange and with ball valve for throttling and isolation, at the pier edge for each product to be transported. The intent is for a loading arm manifold with a separate manual isolation plug valve for each product connection. This will allow simultaneous loading and off-loading of different products, each through a dedicated arm. Provide a double block and bleed plug valve at the point which the line is being stripped. Use the following criteria:

a) Provide each branch line to the pier edge with a manual isolation valve located at the main line. Provide thermal relief valves around isolation and check valves to relieve excessive pressures caused by thermal expansion of liquid trapped between shutoff points. See figure 1.

b) Do not provide a gauge outboard of the hose connection shutoff valve because hose movement will indicate the presence or absence of pressure in the hose.

c) If required, provide one or more loading arms at each station.

d) Provide a liquid-filled pressure gauge for each loading arm, located to be easily read from the operator position. This gauge is provided because the drybreak check valve at the end of the loading arm and the rigid piping will not intuitively indicate the presence or absence of pressure at the loading arm.

e) Provide for venting and draining of the branch lines and loading arm manifolds. Provide for manual venting of the branch lines, connect the vents to the oil waste line, similar to a sanitary vent system to avoid spillage. When pier drain lines cannot be sloped back to the pierhead stripping pumps, a design including separate oil waste drain lines, holding tank and dedicated stripping pump is a viable alternative.

f) Provide segregated handling of multiple products through the loading arms, while allowing easy selection of the products to be transported. Double block and bleed valves can be used for this application.

g) Provide a separate pipe and connection for ballast water or offspec fuel if the size of the facility and level of activity warrants it.

h) Provide each hose handling and loading arm area with fixed spill containment as defined in 33 CFR Part 154.

i) Provide hydraulic shock surge suppressors (if required).

7. PIPING SYSTEMS.

7.1 PIPING ARRANGEMENT. Comply with the following criteria:

a) Where simultaneous deliveries of the same fuel may be made by more than one vessel, size fuel headers and related equipment for the total flow rates of all vessels discharging into the headers. Ensure that flow rates are in accordance with requirements.

b) Place pier piping above the pier deck within a containment area for fueling piers and within a trench on berthing piers. Slope piping toward shore to permit stripping. Use gratings as required to allow access across the piping.

c) Provide flexibility in the piping between the pier and the shore to allow for small movement of the pier relative to the shore. Use a suitable pipe bend or offset configuration, preferably in a horizontal plane, that will allow three-dimensional movement. If vertical bends are used, install vents and drains.

d) Provide flexibility in the piping along the pier to allow for pipe growth due to thermal expansion. Horizontal expansion loops are preferred. In cases where space is tight provide vertical expansion loops or bellows expansion joints where necessary. Where practical provide vertical expansion loops with vents and drains.

e) Include in the pier facilities, pipe manifolds for each fuel type arranged parallel to the face of the pier.

f) Pipe hangers are not allowed.

8. EQUIPMENT DESCRIPTIONS.

8.1 LOADING/OFF-LOADING ARMS. Provide articulated marine loading arms for receiving and shipping fuel cargoes so that the connected vessel can move 15 feet (4.6 mm) forward, 15 feet (4.6 mm) aft, and 10 feet (3 m) off the face of the pier and vertically as caused by loading or off-loading of the vessel and tidal changes, without damage to the arm. Provide a hydraulic power assist system for operating loading arms larger than 8-inch (200 mm) nominal size. Equip the end of the loader to be connected to the ship's manifold with an insulating section, a standard ANSI forged steel flange, and a steel quick coupling device, manually or hydraulically operated. Consider breakaway couplings for locations with strong current.

8.2 FUEL HOSES. Loading/off-loading arms are the preferred method to be used. Provide a facility for storing and protecting the hose as near as practical to the pier if hose is provided in lieu of loading/off-loading arm.

8.3 SUBMARINE FUEL HOSES. Provide submarine fuel hose where offshore moorings are used. Use heavy duty, smooth bore, oil and gasoline, marine cargo, discharge hose rated for a working pressure of not less than 225 psig (1550 kPa) and built-in nipples with Class 300 flanges with stainless steel bolts and Monel nuts. Hoses should be U. S. Coast Guard certified.

8.4 METERS.

Provide a turbine or positive displacement meter for each dispensing outlet that might be used simultaneously. With the approval of the Owner, use portable meters where fueling operations are intermittent. Also consider the use of alternative technologies such as ultrasonic meters. Require temperature compensation feature at each meter used for custody transfer.

8.4.1 METERS – POSITIVE DISPLACEMENT. Require flange-connected, cast steel bodied positive displacement meters of the desired pressure and flow rating for the applicable service requirements. Ensure meter has case drain and register. Provide \1\ meter with temperature compensation and adjustable calibration /1/ when there is custody transfer. Ensure meter accessories are compatible with either the mechanical or electronic support equipment selected. Provide an accuracy of plus or minus 0.5 percent when used for custody transfer. Consult the Owner for requirements for the meter to communicate to a remote location or equipment. Consider the use of a card-operated or key-operated data acquisition system. Cards or keys, as appropriate, are coded to identify the receiver of the fuel and to allow access to the fuel. The quantities taken are transmitted to a data-receiving device by electronic pulse transmitters mounted on each meter, and each transaction is automatically recorded.

8.4.2 METERS – TURBINE. Use flange-connected, steel bodied turbine meters of the desired pressure and flow rating for the applicable service requirement. Provide a flow straightener before turbine meters or provide a straight length of pipe at a minimum of ten pipe diameters upstream and five pipe diameters downstream of all turbine meters, or as required by manufacturer. Ensure meter has case drain and register. Provide meter with temperature compensation and adjustable calibration when there is custody transfer. Ensure all supporting equipment for meter is compatible with the turbine meter selected. Provide an accuracy of plus or minus 0.5 percent when used for custody transfer. Consult the Owner for requirements for the meter to communicate to a remote location or equipment. Consider the use of a card-operated or key-operated data acquisition system. Cards or keys, as appropriate, are coded to identify the receiver of the fuel and to allow access to the fuel. The quantities taken are transmitted to a data-receiving device by electronic pulse transmitters mounted on each meter, and each transaction is automatically recorded.

8.5 STRAINERS. Require a basket strainer to protect centrifugal pumps, unless it precludes meeting the net positive suction head of the pump. Whether or not strainers are installed on the suction side of centrifugal pumps, install a spool piece so that temporary strainers can be installed during startup of the system. Strainers are required on the suction side of all pumps, meters, and receipt filtration. Strainers are not required upstream of issue filter/separators or diaphragm control valves. Also:

a) Use flanged basket strainers constructed of steel and fitted with removable baskets of fine Monel metal or stainless steel mesh with large mesh reinforcements.

b) Unless otherwise specified, provide a fine screen mesh as follows:

	Mesh	Size of Opening
Pump suctions (Centrifugal)	7	0.108 inch (2.74 mm)
Pump suctions (Positive Displacement)	40	0.016 inch (0.40 mm)
Receipt Filtration	40	0.016 inch (0.40 mm)
Meter inlets (unless downstream of a filter/separator)	40	0.016 inch (0.40 mm)

c) In all cases, ensure the effective screen area is not less than three times the cross sectional area of the pipe.

d) Strainers upstream of pump shall be quick opening, single screw type.

e) Provide pressure gauges on both sides of the strainer and a differential type gauge across the strainer.

8.6 SURGE SUPPRESSORS. Every effort should be made to control hydraulic surge or shock to acceptable limits by the design of the piping system rather than by the use of surge suppressors. Where this is not possible or becomes extremely impractical, surge suppressor(s) may be incorporated. Use the diaphragm or bladder type equipped with a top-mounted liquid-filled pressure gauge, wafer-style check valve at the bottom, drain above the check valve, and isolation valve. Provide a needle valve around the check valve to permit controlled bleed back of the surge suppresser without rebounding. Locate surge suppressors as close as possible to the point of shutoff that is expected to cause the shock. Surge suppressors can reduce shock pressure but will not eliminate it entirely. The preferred solution to hydraulic shock is conservative piping design, use of loops, and slow-closing valves. Surge suppressors are strictly a last resort solution and require the approval of the Owner prior to designing into a system.

8.7 VALVES.

8.7.1 MATERIALS OF CONSTRUCTION. Require valves to have carbon steel bodies and bonnets. Do not allow valves with aluminum, cast iron, or bronze materials. Use only API fire-safe valves.

8.7.2 ISOLATION VALVE TYPES.

a) Double Block and Bleed Isolation Valves:

- Use these for separation of product services, on tank shell connections, when piping goes above or below ground, between pier and tank storage, and other locations critical to pressure-testing of piping.

- Plug Valves (Double Block and Bleed): Use double-seated, tapered lift, lockable, plug type valves with a body bleed between the seats (double block and bleed) in critical applications such as separation of product services, when piping goes above or below ground, between pier and tank storage, and other locations critical to pressure-testing of piping. Valves shall be designed so that if the synthetic seating material is burned out in a fire, a metal-to-metal seat will remain to affect closure and comply with API Std 607. Lubricated plug valves are not allowed. Include integral body cavity thermal relief valve.

- Ball Valves (Double Block and Bleed): Use double-seated, trunion mounted, lockable, ball type valves with a body bleed between the seats (double block and bleed). These will be very rarely used but are acceptable as an alternative to double block and bleed plug valves in applications where the valve is operated very infrequently. An example is isolation valves in the middle of piers that are only closed to perform pressure testing of piping. Valves shall be designed so that if the synthetic seating material is burned out in a fire, a metal-to-metal seat will remain to affect closure and comply with API Std 607. Include integral body cavity thermal relief valve.

- Gate Valves (Double Block and Bleed). Use double-seated, lockable, gate type valves with a body bleed between the seats (double block and bleed). These will be very rarely used but are acceptable as an alternative to double block and bleed plug valves and double block and bleed ball valves only when other double block and bleed valves will not physically fit. Valves shall be designed so that if the synthetic seating material is burned out in a fire, a metal-to-metal seat will remain to affect closure and comply with API Std 607. Single seated gate valves are not allowed. Include integral body cavity thermal relief valve.

b) Quick Opening/Frequent Opening Isolation Valves

- Use these for less critical applications where double block and bleed shutoff is not required.

- Ball Valves: Ball type valves may be used as valves for quick or frequent opening applications when a double block and bleed valve is not required. Ball valves shall be designed so that if the synthetic seating material is burned out in a fire, a metal-to-metal seat will remain to affect closure and comply with API Std 607. Use Teflon or Viton synthetic seals or seating material. Use full port ball valves with exact same diameter of the pipe within ten pipe diameters upstream and/or five pipe diameters downstream of a flow or pressure control valve, or a flow-sensing device such as a venturi. Valves should comply with API Std 608.

c) Butterfly Valves: Butterfly valves are not allowed.

d) Use full port valves with exact same diameter of the pipe when line pigging is required.

8.7.3 ISOLATION VALVE OPERATORS. Provide manually operated valves not specified for remote, automatic, or emergency operation. Use geared operators for ball and double block and bleed plug valves larger than 6 inches (150 mm). Double block and bleed gate, ball and double block and bleed valves specified for remote, automatic, or emergency service may have electric motor operators, if approved by the Owner. Provide locking tabs on isolation valves to allow padlock to be used for lock-out during maintenance. Provide chain operators on valves which are located 72 inches (1800 mm) or higher above grade.

8.7.4 ISOLATION VALVE LOCATIONS.

a) Provide an isolation valve on each line at the shore end. For piping used only for receiving fuel, also provide a check valve at the shore end. Use double block and bleed type, which may be motor-operated with remote control. To minimize surge potential, use a slow-closing speed, if possible.

b) Provide double block and bleed isolation valves on the aboveground piping at each barge or tanker off-loading and loading connection.

c) Provide double block and bleed isolation valves near the shoreline of a submerged pipeline to offshore moorings.

d) Provide double block and bleed isolation valves on the inlet and outlet connection of each line strainer, filter/separator, meter, diaphragm control valve, and other equipment that requires periodic servicing. One inlet valve and one outlet valve may be used to isolate more than one piece of adjacent equipment which are connected in series.

e) Provide thermal relief valves around all isolation and check valves to relieve excessive pressures caused by thermal expansion of liquid trapped between shutoff points. See figures 1, 2, 3 and 4.

8.7.5 ISOLATION VALVE PITS. Provide fiberglass or concrete pits with a rolling or hinged cover designed in accordance with the DoD Standard Design AW 78-24-28 for all isolation valves installed in non-traffic areas on underground fuel systems. Design valve pits and valve operators so that the valves can be operated by personnel, without confined space entry.

8.8 OTHER VALVES.

8.8.1 CHECK VALVES. Use check valves to prevent backflow through pumps, branch lines, meters, or other locations where runback or reverse flow must be avoided. Check valves may be of the swing disk, globe, dual plate hinged disk, spring-loaded poppet, ball, or diaphragm-actuated types. Use checks of soft-seated non-slamming type with renewable seats and disks. Ensure check valves conform to API Spec 6D. Use non-surge check diaphragm control valves with flow control feature on the discharge of all pumps. When using non-surge check diaphragm control valves on pump discharge, consider the use of a spring type wafer check before the diaphragm control valve to prevent sudden flow reversals during shutdown from passing back thru the pump before the diaphragm control valve diaphragm chamber is filled and reacts by closing the valve.

8.8.2 THERMAL RELIEF. Provide thermal relief valves around isolation and check valves to relieve excessive pressures caused by thermal expansion of liquid trapped between shutoff points. See figures 1, 2, 3 and 4.

8.9 PRESSURE OR PRESSURE/VACUUM GAUGES. Use glycerin-filled or silicone-filled pressure gauges of range and dial size, as necessary, but not less than 0 to 160 psig (0 to 1100 kPa) pressure range and 4-inch (100 mm) diameter dial. Also:

a) Use pressure gauges upstream and downstream of strainers and filters/separators. A differential pressure gauge may be used in lieu of gauges on each side.

b) Install compound (pressure/vacuum) gauges on the suction side of each pump at fuel storage tanks.

c) Provide a lever handle gauge cock and pressure snubber in each pressure gauge connection.

d) Provide a pressure gauge on each side of the pipeline shutoff valve at the shore end of each pier-mounted pipeline. Provide the indicating pointer with a high-pressure-reading tell-tale indicator suitable for reporting the highest pressure experienced since last reset. Provide for non-contact resetting of the tell-tale by means of a small magnet.

e) Provide a pressure gauge on each branch line at each fueling station on each pier-mounted pipeline. Ensure that the pressure gauge is legible from the fuel hose connection array and from the pantograph loading arm location (if provided).

f) Provide a pressure gauge on each marine loading arm assembly (if provided). Ensure that the gauge is visible by the operator.

g) Pressure gauges shall be installed so that they are testable without removing them from the piping.

8.10 STRIPPER PUMPS. Provide positive displacement stripper pumps for emptying loading arms, hoses, and manifolds. Provide a stripper pump to reclaim each clean product from each main product line, or connect the product lines to the oil waste drain line. Conduct an economic analysis of the two alternatives to determine the appropriate choice. Larger, longer, or more frequently drained lines will favor the stripper pump choice. Use a stripper pump on multi-product lines, but do not exceed acceptable limits of cross contamination. Provide a dedicated stripper pump to each separate product line, such as aviation turbine fuels.

8.11 EXCESS FLOW SENSORS. In piping used for both loading and off-loading, provide a sensor that will alarm both the control room and at the pier to detect excess flow that might occur in the event of a line break.

8.12 SOLID CYCLONIC SEPARATORS. In facilities which receive product by tankers or barge, consider the use of solid separators in the receiving lines as part of pre-filtration to remove gross impurities from the incoming product. In systems equipped with filter/separators in the receiving lines, locate strainers or cyclonic separators upstream of the filter/separator. Ensure that there is no slug valve feature on the filter/separator. Consider the use of automatic water drains. Do not allow reverse flow thru cyclonic separators.

8.13 GROUNDING SYSTEMS. Provide grounding systems for barges.

9. PRODUCT RECOVERY SYSTEMS. Provide a product recovery system to collect and store usable aviation turbine fuel that would otherwise become waste from operational or maintenance activities. Consider a product recovery system for other products.

10. WEATHER SHEDS. Provide adequate shelter for personnel, as well as for spill containment booms, absorbent material, and other weather-sensitive equipment.

11. CANOPIES. Provide a canopy, as directed by the Owner, for all aboveground equipment including pumps, meters, strainers, filters, control panels, electrical panels, and motor control centers (MCCs).

12. SPECIAL CALCULATIONS. Calculate pipeline filling/venting times and draining/stripping times. The larger and the longer the pipeline, the greater the volume of fuel required to fill the line and, therefore, the greater the volume of air required to be vented. Undersized vent lines will delay filling the lines and delay changeover of products in multiproduct lines. Size the vent lines to allow filling of the line at not more than four times the design transit time of the line. Connect vent line to the drain line to avoid spills to the environment. Check vent line air velocity, which must not exceed the allowable air velocity to avoid electrostatic buildup, in accordance with API RP 2003. Vent rate must be not less than the lowest allowable pumping rate from ship or shore. Vent rate must be less than the design transit velocity to minimize hydraulic shock.

13. SAFETY SHOWERS AND EYEWASH FOUNTAINS. Provide manual shutoff valves on the potable water branch to the safety shower and eyewash fountain. Provide a means to seal shutoff valve in the open position. This will ensure operation in an emergency, yet allow for servicing a single shower without shutting off potable water to the whole pier. Design for freeze protection in climates subject to freezing. Install safety showers and eyewash fountains in accordance with ISEA Z358.1.

14. TRAFFIC BOLLARDS. Provide traffic bollards to protect fueling piping and equipment on piers and wharves. Utilize concrete-filled steel pipe of minimum 4-inch (100 mm) diameter and 4-foot (1.2 m) height, embedded in concrete or welded to a steel plate mounted on the structure.

15. SPECIAL DRAINAGE FOR FUELING PIERS.

a) Provide an intercept system to collect oil spills. Place pipes on piers in a curb containment area with a drain system independent of the deck drainage. Provide containment also for loading arms and risers. Provide locking valves in normally closed positions on all containment areas along with sump pumps or other means of removing the spilled fuel to a collection point or tank.

b) In cases where the stormwater collected in the intercept system is contaminated, the water/fuel mixture should be treated as an oil spill as described previously.

16. BALLAST TREATMENT AND SLUDGE REMOVAL.

16.1 BALLAST RECEIVING AND TREATMENT FACILITIES.

16.1.1 DESIGN REQUIREMENTS. It is often policy that there should be no discharge of oil or hazardous substances into or upon the navigable waters of the United States, adjoining shorelines, or into or upon the waters of the contiguous zone. Petroleum fuel facilities, which transfer fuel by barge or tanker or which fuel large ships, require ballast water collection and treatment facilities to receive and treat oily ballast from cargo or fuel tanks. Also:

a) Blend the fuel oil which has been reclaimed from the ballast water during the collection and treatment process with boiler fuel oil for use in shoreside boilers. Perform a quality assurance check on the reclaimed fuel oil to ensure that it meets the minimum requirements for shoreside boiler fuel. Dispose of sludge accumulated during the collection and treatment of ballast water in accordance with applicable hazardous waste management disposal procedures.

b) Select and design the appropriate treatment system based on an evaluation of the types of oil/water mixtures that may be encountered at the particular facility. If possible, base the evaluation on samples of typical ballast water receipts and tank washings including the following:

- Whether they are simple mixtures, simple gravity suspensions, or chemically stable emulsions.
- The specific gravity and viscosity of the oil in the mixture.
- Whether other substances, such as chemicals or bacteria, in the mixtures must be removed.
- The general condition of the ship's tanks expected to be discharged (e.g., new, clean, coated, well maintained, or dirty and normally full of sludge, scale, and rust).

- Whether ballast water is clean sea water or polluted harbor water.
- Whether the treatment system proposed ("ship's waste off-load barge" or fixed shore-based facilities) meets the standards of effluent water quality established by local environmental regulations.

c) If it is determined that both simple mixtures and emulsions are present, consider the possibility of using two segregated separate systems, one for gravity separation and the other for breaking emulsions. Avoid mixing the two types of suspensions when possible.

16.1.2 RECEIVING AND SETTLING TANKS. The minimal ballast water receiving facility usually requires two storage tanks, usually of equal capacity, to be used alternately as receiving and settling tanks. If these tanks are sized to allow 4 to 5 days undisturbed settlement, separation of simple suspensions of light oils in water can be achieved. Provide the following fittings and appurtenances:

a) An automatic float gauge suitable for use with transmitting device for remote readout.
b) One cable-operated swing-line assembly on the oil outlet pipe.
c) One shell fill nozzle.
d) Valved sample connections in the shell, having nonfreezing-type valves in cold climates, every 2 feet (0.6 m) vertically, easily accessible from the ladder or stairway.
e) When chemical feed is provided, a chemical feed inlet valve, to be nonfreezing type in cold climates.
f) When air blowing is provided, a perforated pipe air sparger for mixing. Make the perforations in the sides of the pipe to avoid plugging by settling solids. Use nonfreezing-type air inlet valve(s) in cold climates.
g) Sight glass or look box on oil outlet line.
h) Sight glass or look box on water outlet line.
i) Oil sump tank with high-level alarm.
j) Water and oil pumps as required to move fluids from receiving tanks or from oil sump tanks. For transfer of oily water, use low-speed-type pumps to minimize emulsification.

k) If heaters are required to reduce oil viscosity and promote separation, use either tank wall heaters or internal pipes. Keep internal pipes at least 2 feet (0.6 m) above the tank floor.

l) Insulation for tanks that will be regularly heated.

m) Provide automatic temperature controls and thermometers for all heated tanks.

16.1.3 OIL/WATER SEPARATORS. Separate water/fuel mixtures from storage or settling tanks with an API oil/water separator. Recycle the fuel portion and pass the water portion to another treatment process. Do not discharge water drawn from tanks to surface water without additional treatment and permits.

17. SLUDGE REMOVAL SYSTEMS.

17.1 DESIGN REQUIREMENTS. Install sludge removal systems where the accumulation of sludge in substantial quantities is likely to occur on a regular basis. Sources of such sludge are a ballast water treatment system, a contaminated fuel recovery system, or frequent cleaning of shore or ships' tanks. If routine cleaning of clean product storage tanks occurs on an irregular basis, sludge removal systems are not required.

17.2 SLUDGE DISPOSAL.

a) Where possible, provide pumps, tanks, and piping to return sludge containing recoverable oil to the contaminated oil recovery system. If this is not possible, consider transferring the sludge to a refinery or waste oil treatment facility.

b) Provide a tank or tanks with transfer pump(s) for pumpable sludges that are unreclaimable. Include piping for receiving sludge and for mixing other low viscosity waste oils for thinning as required. Ensure that tanks are dike-enclosed and have cone bottoms.

c) Provide tank heating where climate conditions prove necessary.

d) Coordinate sludge disposal method and design with facility environmental office.

e) Enclose the sludge disposal facility with a security fence to prevent unauthorized entry. Do not use this facility for disposal of sand, gravel, rust scale, or other solid nonpumpable matter found on tank bottoms.

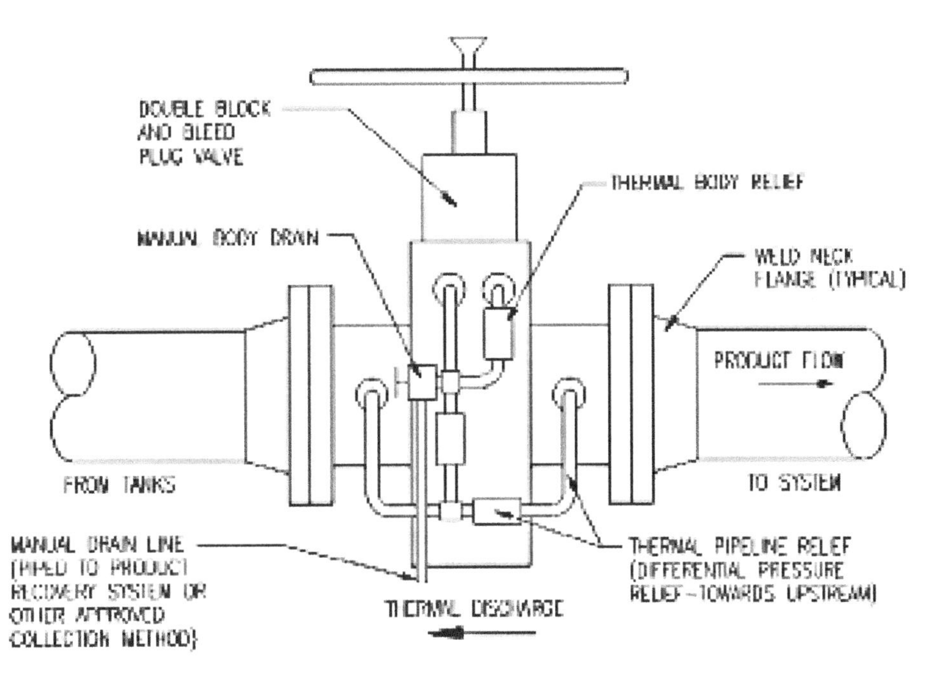

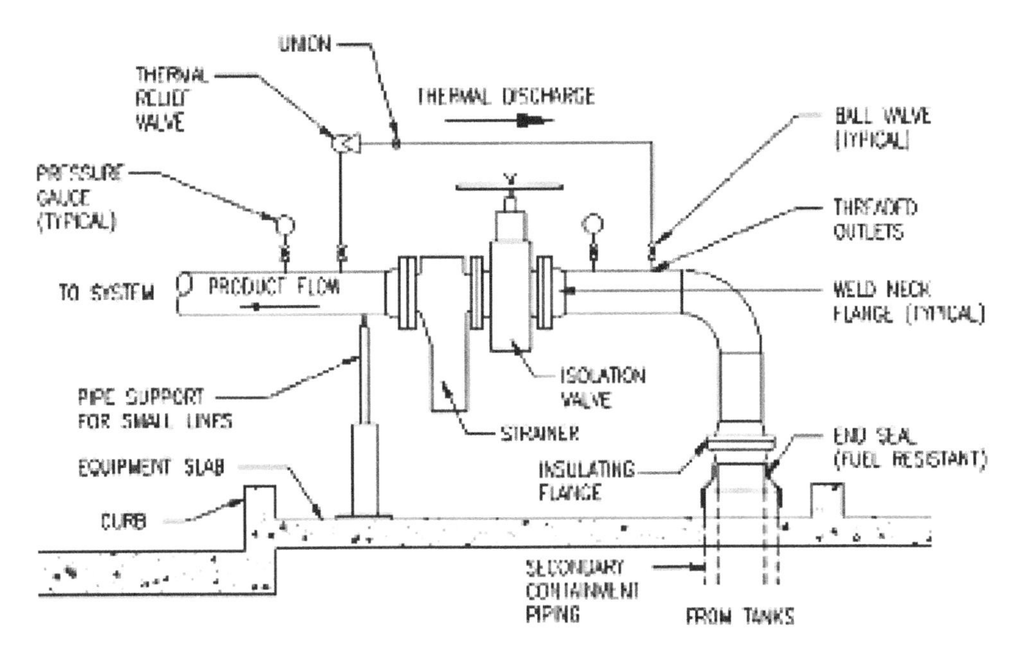

Figure 1

Thermal Relief Piping Systems Integral Valve and Conventional

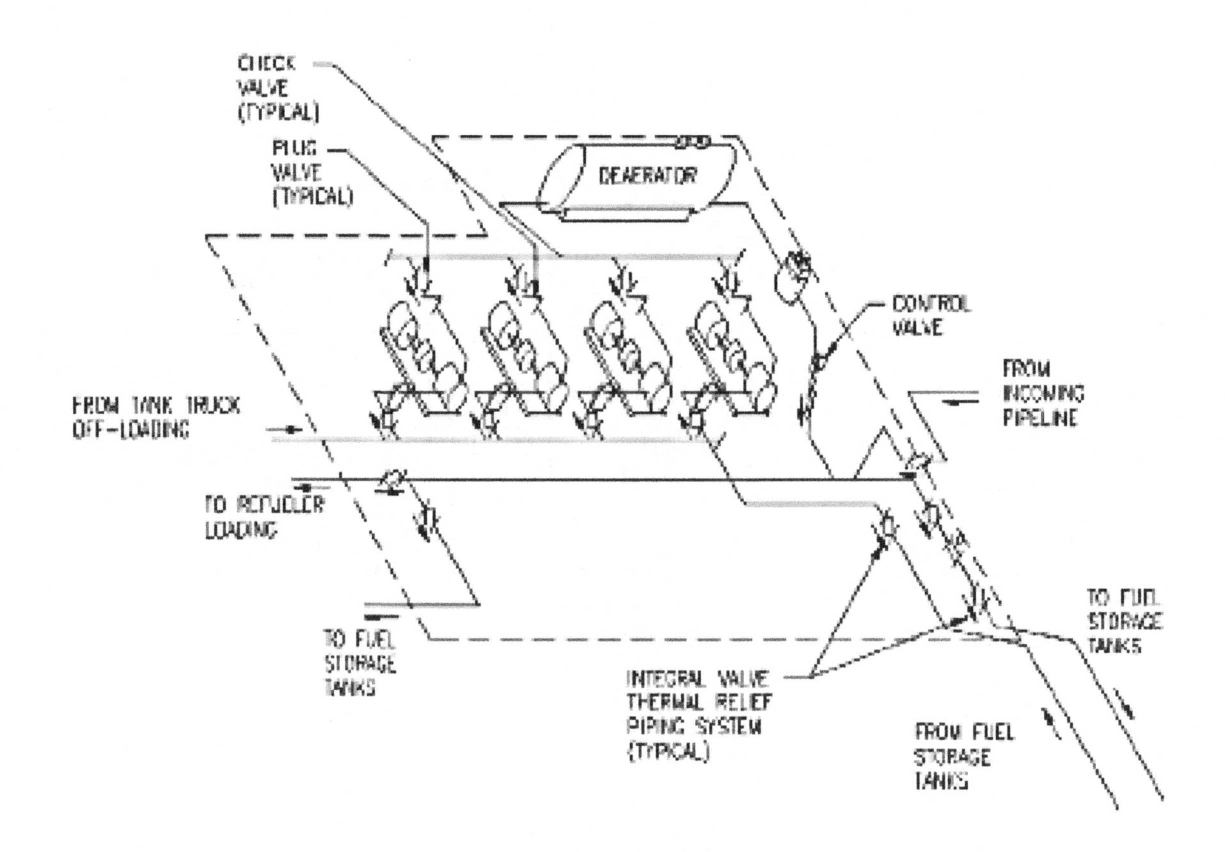

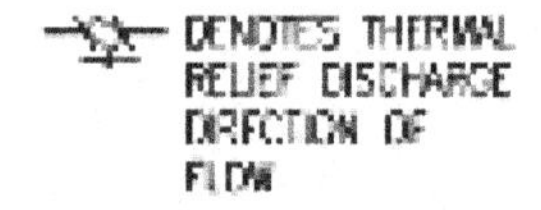

EQUIPMENT PUMPHOUSE OR PADS

NOT TO SCALE

Figure 2

Thermal Relief Piping Systems Equipment Pump House or Pads

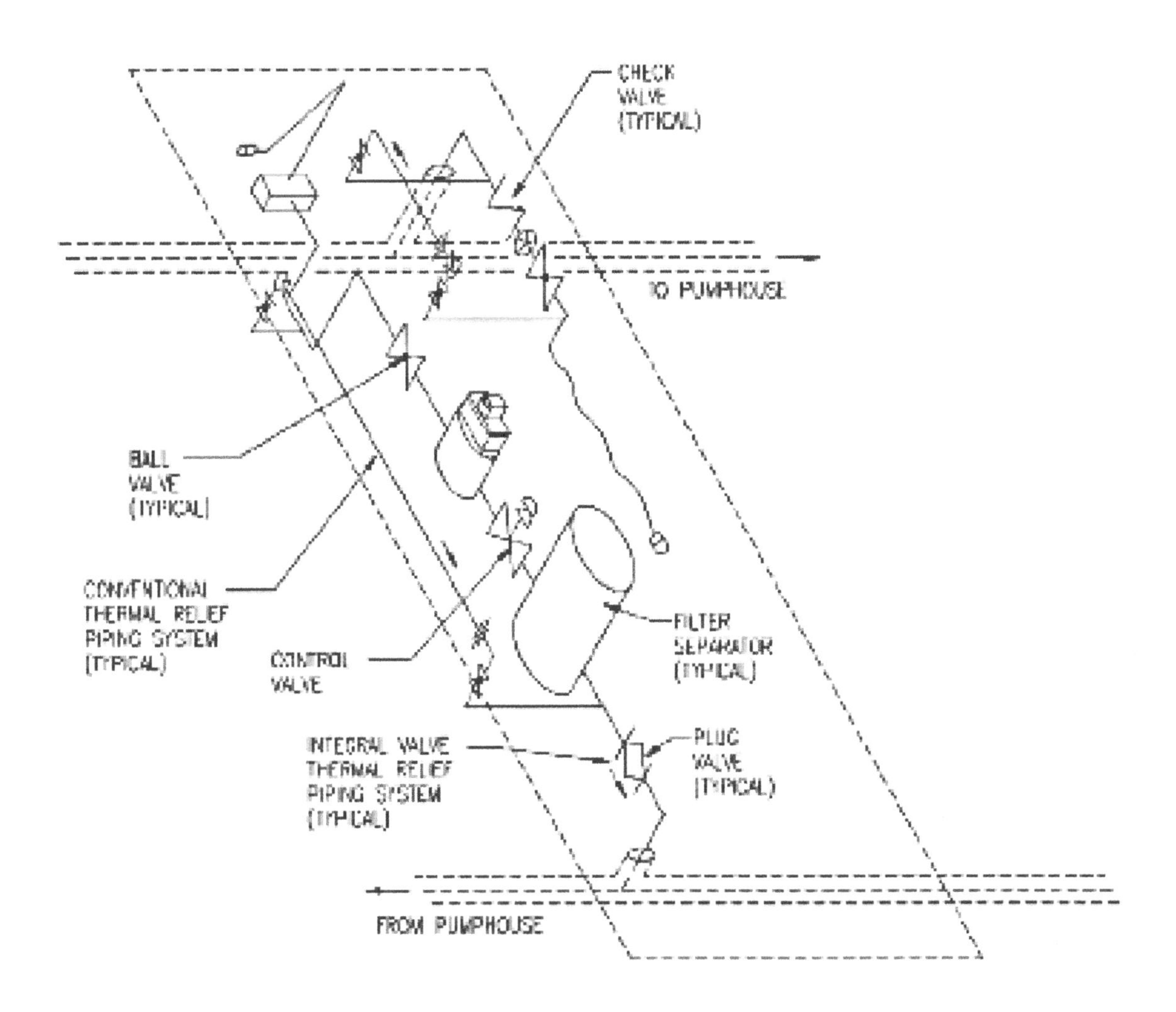

TANK TRUCK AND REFUELER EQUIPMENT RACKS

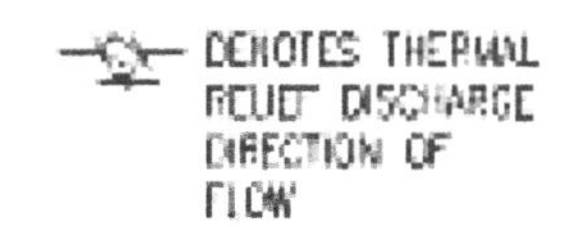

NOT TO SCALE

Figure 3

Thermal Relief Piping Systems Tank Truck and Refueler Racks

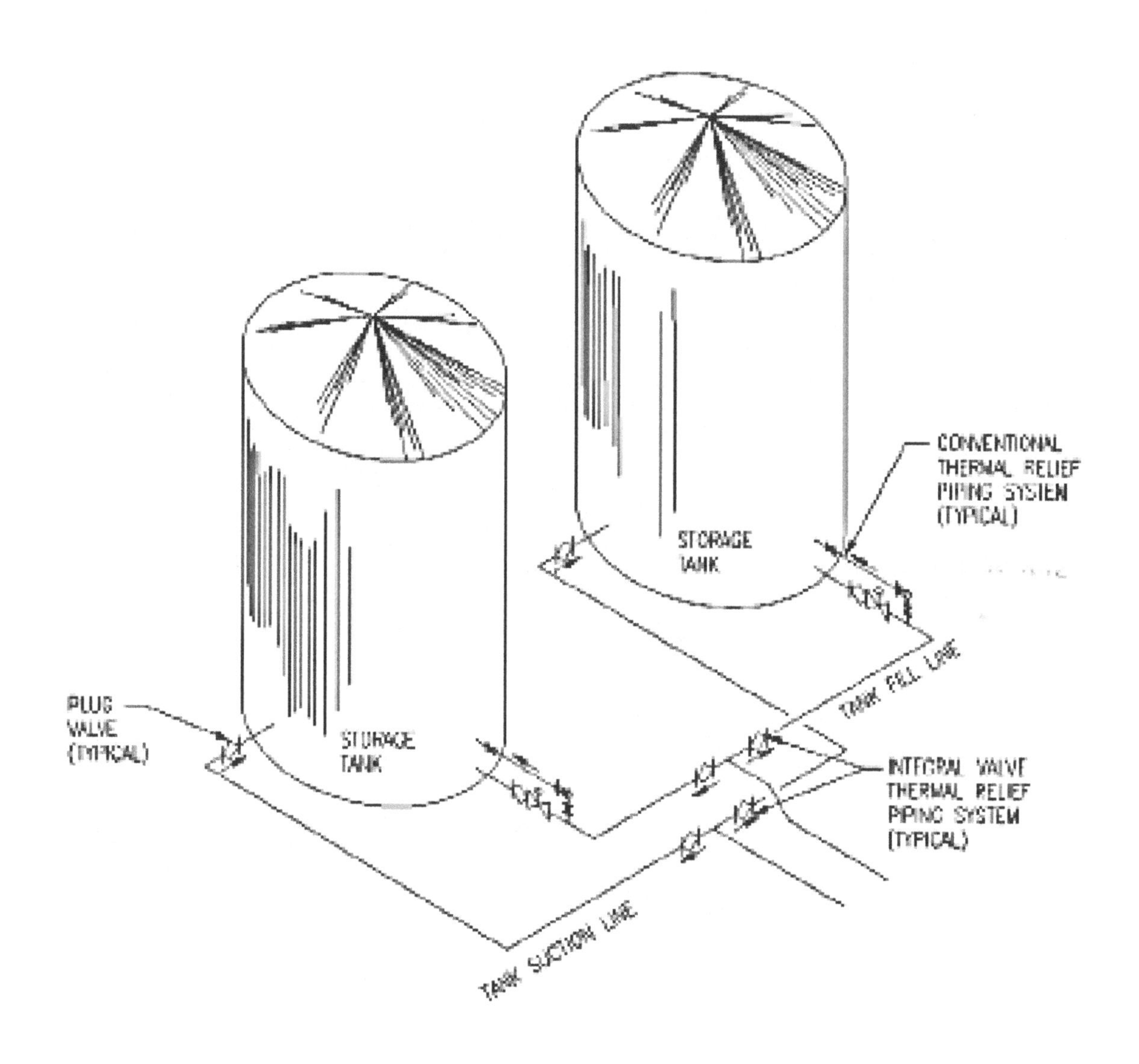

NOT TO SCALE

Figure 4

Thermal Relief Piping Systems Storage Tanks

Made in the USA
Las Vegas, NV
26 May 2024